U.S. INDIVIDUAL FEDERAL INCOME TAXATION: HISTORICAL, CONTEMPORARY, AND PROSPECTIVE POLICY ISSUES

STUDIES IN MANAGERIAL AND FINANCIAL ACCOUNTING

Series Editor: Marc J. Epstein

STUDIES IN MANAGERIAL AND FINANCIAL ACCOUNTING
VOLUME 11

U.S. INDIVIDUAL FEDERAL INCOME TAXATION: HISTORICAL, CONTEMPORARY, AND PROSPECTIVE POLICY ISSUES

BY

ANTHONY J. CATALDO II

*Department of Accountancy, Western Michigan University and
Department of Accounting and Finance, Oakland University
Michigan, USA*

ARLINE A. SAVAGE

*Department of Accounting and Finance, Oakland University,
Michigan, USA*

2001

JAI
An Imprint of Elsevier Science

Amsterdam – London – New York – Oxford – Paris – Shannon – Tokyo

ELSEVIER SCIENCE Ltd
The Boulevard, Langford Lane
Kidlington, Oxford OX5 1GB, UK

First edition 2001

Library of Congress Cataloging in Publication Data
A catalog record from the Library of Congress has been applied for.

British Library Cataloguing in Publication Data
A catalogue record from the British Library has been applied for.

ISBN: 0-7623-0785-4

∞ The paper used in this publication meets the requirements of ANSI/NISO Z39.48-1992 (Permanence of Paper).
Printed in The Netherlands.

CONTENTS

PART 3. CAPITAL GAIN/LOSS TAXATION

PART 4. GENERATIONAL TAXATION

LIST OF FIGURES AND TABLES

PART 1. INTRODUCTION TO U.S. INDIVIDUAL TAXATION

PART 2. FAMILY TAXATION

PART 3. CAPITAL GAIN/LOSS TAXATION

PART 4. GENERATIONAL TAXATION

PART 5. PROSPECTIVE INDIVIDUAL TAX POLICY ISSUES

PREFACE

Much of the academic research and literature dealing with individual taxation has focused on *tax complexity* (the *ability* to comply with increasingly complex U.S. Federal income tax legislation) and *tax compliance* (the *desire* to comply with U.S. Federal income tax legislation) issues. Researchers of these topics have generally excluded *generational accounting* (Kotlikoff, 1992) and related *tax equity* topics.

In Crum (1992, 19–20), Frances Ayres suggested that six *individual* income tax issues warranting future investigation should include: (1) effects on stock returns, (2) revenue impacts and (3) behavioral responses to tax law changes, (4) distribution of the tax burden, (5) costs and solutions to non-compliance, and (6) impacts of (non-income) taxes on individual taxpayers. We have designed this monograph as a source of reference for those interested in investigating these and other individual Federal income tax policy and related research issues. Coverage includes our thoughts on the future directions of individual Federal income tax policy and the topics likely to be of greatest interest to academic researchers and policy makers.

Shoven and Whalley (1992) assembled and edited a selection of studies on contemporary U.S.-Canadian tax policy and tax harmonization efforts. We believe that historical patterns and trends in U.S.-Canadian tax policy formulation are useful in gaining insights into future policy formulation. Therefore, to the extent practicable, we have included selected descriptive statistics, discussions and references to U.S.-Canadian similarities and differences throughout this monograph.

ACKNOWLEDGEMENTS

Both of our institutions provided financial support for the production of this monograph. This research was partially supported by a 2000 spring/summer Haworth College of Business research grant from Western Michigan University in Kalamazoo, Michigan, and a 2000 School of Business Administration spring/summer research grant from Oakland University in Rochester Hills, Michigan.

We are very grateful to Marc Epstein, Susan Oppenheim and Sammye Haigh for believing in this project and supporting us through the delays resulting from an extended U.S. Presidential election (2000). We are also very grateful to David Hurtt (Western Michigan University) and David Rozelle (Western Michigan University) for their thoughtful reviews of an earlier draft of this monograph.

Any errors remain our responsibility.

PART 1

INTRODUCTION TO U.S.
INDIVIDUAL TAXATION

CHAPTER 1

INTRODUCTION

(The) tax system is more than a revenue-raising device, it is an instrument of national policy (Heller, 1969, as cited in Pollack, 1996, 13)

Tax law and the U.S. federal income tax (FIT) consist of individual and corporate income taxes, estate and gift taxes, and excise and customs taxes. Individual federal income taxes approximated 50% of all federal tax revenues for the 1996 tax year (IRS SOI, 1996). Corporations provided only 13% for the same period. Yet, much of the contemporary academic research in the accounting literature has focused on *corporate* tax topics. This monograph focuses entirely on topics relating to U.S. *individual* federal income taxation.

This monograph was motivated by the desire to serve three primary purposes:

First, we provide *historical* data (1913–), and references to the source of this information, on topics of contemporary interest. Therefore, this monograph will serve as a source of reference for academics, graduate students, and those interested in tax policy and the historical evolution of selected contemporary individual federal income tax issues.

Second, we will extend this historical analysis to focus on selected issues addressed in *contemporary* research. We operationally define the contemporary period as the past quarter of a century (1976–2000).

Third, we attempt to identify any discernible patterns and/or trends in individual federal income taxation for the next few decades. We expect the driving force for tax policy changes to evolve from: (1) changing U.S. demographics and, (2) the realization of anticipated U.S. budget surpluses. From this, our objective is to identify significant *prospective* topics likely to warrant further development and examination (2000–).

In summary, our objective is to examine historical and contemporary trends or patterns in U.S. individual federal income taxation to provide direction for future tax policy-based research. To achieve this, we have made extensive use

3

of the annual *Statistics of Income* (SOI) publications produced by the Internal
Revenue Service (IRS) for the development of U.S. descriptive statistics. It is
the practice of the IRS SOI to organize individual tax return data by adjusted
gross income (AGI). The U.S. summary data contained in this monograph will
also use AGI-based measures for many of the percentage and tabular analyses.

U.S.-CANADIAN TAX POLICY COMPARISONS

Many similarities exist between the evolution of U.S. (1913–) and Canadian
(1917–) individual federal income tax. For example, in the historical period,
particularly that leading up to World War II (WWII), Canadian tax policy
formulation was similar to and appeared to follow that of the U.S. However,
during the more contemporary or post-WWII period, Canadian tax policy
changes and formulation has preceded the U.S. in the replacement of tax
deductions with tax credits, and particularly in the use of child care and
individual tax credits. These changes in Canadian tax policy may influence
U.S. tax policy and are consistent with the goals of tax harmonization between
the two countries (Shoven & Whalley, 1992). Therefore, our examination of
U.S. individual federal income tax policy includes selected discussions,
analyses and use of the Canadian counterpart to the U.S. IRS SOI, *Taxation
Statistics* (TS), published by the Statistical Services Division of Revenue
Canada (RC).

It is the practice of RC to organize aggregate individual tax return data by
"total income assessed", a measure comparable to U.S. total income (TI). For
comparative purposes, the summary data contained in this monograph will also
use "total income assessed"-based measures for many of the percentage and
tabular analyses.

From these statistical publications, we have developed longitudinal data for
both the U.S. and Canada, for quick access and use by others, and for
regression model development and hypothesis development and testing. It is
our hope that those requiring additional or more detailed data may find this
monograph useful as a reference to potential sources of both U.S. and Canadian
data and studies of individual federal income tax policy issues.

ORGANIZATION OF THIS MONOGRAPH

The organization of this monograph is provided in Fig. 1-1.

Part One provides an introduction to individual taxation. The appendices to
Chapter 2 include a review of the basic components of U.S. individual income
taxation (Appendix 2A) and a contemporary U.S./Canadian individual tax

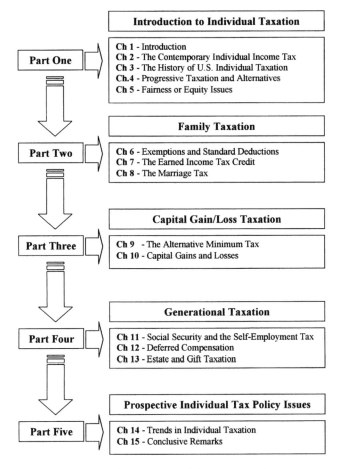

Fig. 1-1. Organization of this Monograph.

return format comparison (Appendix 2B). Where relevant and practicable, U.S./Canadian comparisons are made throughout this monograph. Chapter 3 continues this introduction, by providing for a brief history of post-16th Amendment-based U.S. individual taxation. Topics not covered extensively in later chapters are addressed, from a historical or evolutionary perspective, in the appendices to Chapter 3. These include deductible charitable contributions (Appendix 3C), medical and dental expenses (Appendix 3D), income-averaging (Appendix 3E), the maximum tax (Appendix 3F), all five categories of deductible and non-deductible interest expenses (Appendix 3G), the two-earner married couple deduction (Appendix 3H), and passive activity losses (Appendix 3I). Appendix 3H is particularly important, as the two-earner married couple deduction was the mechanism proposed by President-elect (and, now, President) Bush to reduce the marriage tax penalty. Chapter 4 examines progressive U.S. taxation, discusses the Canadian income tax and national sales tax, and includes some discussion of the alternatives suggested for the U.S. in recent decades. Included in the appendices to Chapter 4 are examinations of historical trends toward greater progressivity, the move from a *class* tax to a *mass* tax, and the flattening of progressive rates in recent decades (Appendix 4J). Internet business-to-consumer sales taxation issues are discussed in Appendix 4K. Chapter 5 reviews the principle of *fairness*, horizontal and vertical equity techniques, strengths and weaknesses, and the concept of explicit and implicit taxes. Appendix 5L provides a comparison of U.S. and Canadian inflation-indexing, a process designed to promote equity through the elimination of *tax bracket creep*.

Part Two examines issues relating to *family taxation*. Chapter 6 summarizes the U.S. history of the personal exemption and the filing status-based standard deduction. A Canadian historical counterpart is provided for comparative purposes. Chapter 7 follows with an examination of trends emerging from the re-establishment and growing importance of the earned income (tax) credit (1975–). This chapter summarizes the mechanics, strengths, weaknesses and ongoing administrative problems with this "workfare" feature of the U.S. individual income tax system. Chapter 8 concludes with an examination of the inequities associated with the marriage tax penalty or bonus. Again, comparisons are made to the Canadian system. A contemporary analysis of *marriage-to-divorce ratios* is made between the two countries, and the differences and similarities are discussed. Appendix 8M briefly summarizes the international history of family policy and taxation in selected countries.

Part Three summarizes key issues relating to the taxation of *capital gains and losses*. Chapter 9 traces the historical evolution of the contemporary alternative minimum tax and the concept of "tax preferences". Historically, two alternative taxes were designed to provide alternative *reductions* to the regular tax rate on capital gains. The emphasis shifted as capital gains received preferential treatment, and the minimum tax and the contemporary alternative minimum tax evolved. Chapter 10 provides a more direct examination of the historical and contemporary trends relating to the taxation of capital gains and losses. The fundamental issue of fairness associated with the taxation of capital gains is addressed and the history of holding periods (for preferential treatment) is provided. Appendix 10N examines capital gains in the context of the contemporary day-trading phenomenon.

Part Four examines the issue of *intergenerational taxation*. Beginning with a historical perspective, Chapter 11 examines Social Security and the taxation of the self-employed. Social Security represents an intergenerational contract for the funding and distribution of retirement, old-age, and medical coverage. The chapter includes a novel contemporary investigation into the impact of contemporary self-employment tax increases on small business formation and profitability. Chapter 12 focuses on the establishment and growth of tax deferred compensation. Comparisons to Canadian counterparts are provided. Chapter 13 concludes this section with a review of estate and gift taxation. Again, Canadian comparisons are provided.

Part Five contains our summary of *prospective individual tax policy issues*. In Chapter 14, we examine the historical, contemporary, and prospective trends in U.S. individual taxation. We expect changing American demographics and anticipated U.S. budget surpluses to drive future tax policy. With these underlying forces in mind, we conclude with suggestions for future academic research and public policy recommendations in Chapter 15.

Appendix I – Tables and Figures for Main Chapters

In addition to the tables and figures contained in the body of the 15 main chapters, Appendix I contains many supplemental tables and figures useful for historical and contemporary reference.

We have included these tables and figures as an efficient means of providing the reader with a variety of supplemental information. Therefore, a question arising from a reading of the main body of this monograph might be answered, or selected as a topic for more detailed investigation through the many sources cited in the footnotes and references sections, after a review of these supplemental tables and figures. The graduate student or academic researcher might find this information useful for preliminary statistical analysis (e.g. regression formula development).

Appendix II – Supplemental Chapters

Appendix II contains additional, more narrowly focused discussions on topics introduced in the 15 main chapters. For example, Appendix 3H more fully develops historical information relating to the two-earner married couple deduction. This topic is first introduced in Chapter 3. The resurrection of this "adjustment to income" has both historical and contemporary relevance, as it was included in President Bush's 2000 campaign platform and his early-2001 tax proposal.

Appendix III – Table and Figures for Appendix II (Supplemental Chapters)

Finally, just as Appendix I contains supplemental tables and figures relating to the 15 main chapters, Appendix III contains supplemental tables and figures relating to the shorter, more narrowly focused discussions contained in Appendix II.

CHAPTER 2
THE CONTEMPORARY INDIVIDUAL FEDERAL INCOME TAX

> ...the current tax system is a complex hybrid between an income and consumption tax, with some features that violate the principles of either system (Aaron & Gale, 1996, ix).

INTRODUCTION

U.S. Federal income taxes are a combination of individual income taxes and social security taxes. The contemporary U.S. individual federal income tax system is *progressive*. Therefore, the progressive nature of the U.S. individual federal income tax system is addressed in this chapter. The Medicare component of the Federal Insurance Contributions Act (FICA) tax or the Self-Employment Contributions Act (SECA) tax currently takes the form of a *flat* (or *regressive*) tax. The Old-Age, Survivors, and Disability Insurance (OASDI)[1] or Social Security components of FICA and SECA taxes are also *flat* (but, from the perspective of the *high-income* taxpayer, FICA and SECA taxes may also be viewed as *regressive*). Together, these taxes accounted for approximately 82% of all U.S. federal income taxes collected for the 1996 tax year (IRS, 1996).

Figure 2-1 graphically illustrates the *progressive* nature of the contemporary U.S. individual federal income tax, which is applied to *taxable income*. Tax rates or brackets range from a low of 15% to a maximum rate of 39.6%. As *taxable income* increases, the federal income tax rate applied to the last taxable dollar also increases.

Figure 2-2 graphically illustrates the *regressive* or *flat* nature of the U.S. Social Security tax, which is applied to the U.S. individual taxpayer's *earned income*. Both FICA (6.2%) and Medicare (1.45%) taxes are applied on all earned income through a maximum *wage base* or *ceiling*. This ceiling is inflation-indexed ($76,200 for 2000 tax year). For taxpayers with earned

9

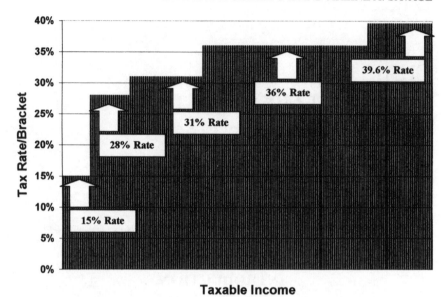

Fig. 2-1. The U.S. *Progressive* Federal Income Tax[a] (2000).

[a] This illustration of all five tax brackets is for a single taxpayer.

income above this ceiling, the Medicare (*flat*) tax component continues to apply beyond the wage base and without a ceiling.

The progressive nature of the U.S. individual federal income tax system has been modified in recent decades. In an effort to achieve greater *horizontal* and *vertical equity* (discussed in Chapter 5), tax credits have been established for low- to moderate-income taxpayers, and deduction exclusions, personal and itemized deduction phase-outs or ceilings have been established to limit federal income tax savings for high-income taxpayers.

A summary of the basic components of the contemporary U.S. individual federal income tax is provided in Appendix 2A. This appendix is intended for those less familiar with the basic U.S. individual income tax return format.

SIMPLIFICATION OF THE U.S. INDIVIDUAL FEDERAL INCOME TAX

The *Tax Advisor*, a publication of the American Institute of Certified Public Accountants (AICPA) included in its recommendations the simplification of the following six areas relating to individual taxation (Laffie, 1998):

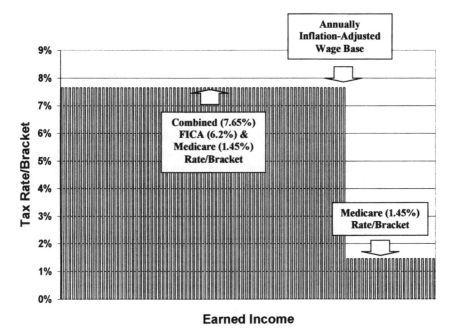

Fig. 2-2. The U.S. *Flat/Regressive* Social Security (6.2%) and *Flat* Medicare (1.45%) Tax (2000).

(1) Income level-based phase-outs (see Chapter 2);
(2) The earned income (tax) credit (EI(T)C; see Chapter 7);
(3) The marriage tax penalty (MTP; see Chapter 8);
(4) The alternative minimum tax (AMT; see Chapter 9);
(5) The capital gains tax (CG; see Chapter 10); and
(6) Deductions for self-employed (SE) health insurance (see Chapter 11).

We examine these areas in the chapters indicated above.

The so-called (1) "kiddie tax" and (2) "safe harbors", dealing with the timing and minimum percentage requirements for contemporary quarterly estimated tax payments, (3) the legal and operational definitions of "employee" and "independent contractor", and (4) "half-year" requirements were also recommended for simplification by the AICPA. However, these topics do not represent individual tax *policy* issues, are too narrowly focused, and are beyond the scope of this monograph.

U.S.-CANADIAN COMPARISONS

Historically, many similarities have existed between the U.S. (1913–) and Canadian (1917–) systems of individual income taxation. Over the past two decades, some important differences have evolved. Where appropriate, historical and contemporary Canadian comparisons are made to provide insights into the U.S. counterparts. To a much lesser degree, historical and contemporary international comparisons are made for the same reason.

A summary of the basic components of the contemporary Canadian individual federal income tax is provided in Appendix 2B. This appendix is intended to introduce the basic framework for the contemporary Canadian individual income tax return.

CHAPTER 3
THE HISTORY OF U.S. INDIVIDUAL TAXATION

The structure of the federal income tax system is under increasing attack from the public, whose case has been taken up by politicians, economists and tax specialists, among others. Many believe that radical reform is needed to prevent further deterioration of the system since piecemeal attempts at tax reform by simplification and increased compliance efforts are seen as being ineffective (Phillips & Previts, 1983, 64).

INTRODUCTION

This chapter provides a broad introduction and review of the history of the contemporary U.S. individual federal income tax system. In this chapter, applicable revenue or tax acts are emphasized in italicized bold type. The major points of these acts are summarized. However, in cases where the tax laws have undergone multiple changes, were transitory, and/or are further investigated in later appendices or chapters, the appendix or chapter reference is given.

FAILED EARLY ATTEMPTS AT INDIVIDUAL FEDERAL INCOME TAXATION

Income taxes were first levied to finance the Civil War (1862–1871),[2] and then again in 1894.[3] Both efforts were later declared unconstitutional and repealed as direct taxation without apportionment.

The present U.S. system of individual federal income taxation began with the *Corporation Excise Tax Act of 1909*,[4] the constitutionality of which was sustained as a *business privilege* or *excise* tax and not an *income* tax. Congress adopted the 16th Amendment to the Constitution,[5] providing for the valid imposition of an *individual* federal income tax without apportionment.

The *Tariff Act of October 3, 1913*,[6] which provided for the taxation of individuals as well as corporations, applied only to income earned on and after March 1, 1913. All income earned or accrued prior to that date was tax-free.

THE POST-16TH AMENDMENT PERIOD

The IRS began the publication of descriptive statistics (*Statistics of Income* or SOI) from individual income tax returns in 1916. A later edition of the IRS SOI (1965, 199) provided a framework for the comparison of the summary data compiled in the IRS *Statistics of Income* publications. This framework (adapted) is presented in Fig. 3-1.

This chapter follows the IRS SOI framework as shown in Fig. 3-1. The first period (1913–1943) and second period (1944–1953) both used the *net income* measure/concept. The second period and third period (1954–) both used *adjusted gross income* (AGI) measures, which persists today, and is combined with the contemporary *taxable income* (1954–) measure.

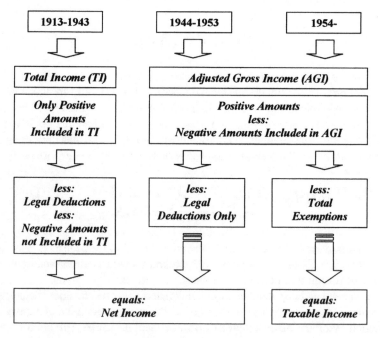

Fig. 3-1. Definitions Used in the Internal Revenue Service *Statistics of Income* Publications (1913–1965).

The First Period of Individual Taxation (1913–1943)

The ***Revenue Act of 1913***[7] exposed taxpayers to a *normal flat* tax of 1% and a *progressive surtax* ranging from 1% to 6%. The normal tax was applied to *taxable income*. The surtax was applied to *net income*. This act was effective for the 1913, 1914,[8] and 1915 tax years.

Taxable income, as operationally defined for this early period of individual federal income taxation, is comparable to the contemporary definition of taxable income, the computation of which is illustrated in Eq. (3-1), as follows:

$$TI = NI - (D + E + PT) \qquad (3\text{-}1)$$

where TI = taxable income, NI = net income, D = dividends, E = exemptions, and PT = pre-taxed items. The dividends (D) on stocks or from the earnings of U.S. domestic corporations were not subject to the normal tax during the initial years of individual federal income taxation. Pre-taxed (PT) items were those income items from which normal "taxes at the source" had already been withheld.

Progressive *surtax* rates were applied to net income (NI), the computation of which is illustrated in Eq. (3-1'), as follows:

$$NI = GI - GD \qquad (3\text{-}1')$$

where GI = gross income and GD = general deductions. The general deductions subtracted in computing *net income* were comparable to our contemporary itemized deductions.

These early individual federal income tax returns were due on or before March 1st of the next calendar year. Full payment (in arrears) of the entire individual federal income tax was due on or before June 30th of the next calendar year.

The ***Revenue Act of 1916***,[9] applicable only for the 1916 tax year, doubled the *normal* tax rate (to 2%) and more than doubled the range of the *surtax* (to a maximum rate of 13%) (IRS SOI, 1926, 38–9 and Blakely, 1916, 839–40). This Act also shortened the time for the payment of the individual federal income tax by two weeks, from June 30th to June 15th of the next calendar year. The estate tax, discussed in Chapter 13, was re-introduced by this Act.

The ***Revenue Act of 1917*** and the ***War Revenue Act of 1917***[10] imposed taxes on incomes previously too small to be taxed (by decreasing exemptions) and increased taxes for those previously subject to the individual federal income tax. Normal taxes were increased and the range of surtaxes was increased to a maximum of 63%.

Beginning with the 1917 tax year, individual taxpayers were permitted a *general deduction*, a category of deductions comparable to the contemporary *itemized deduction*, for charitable contributions. Taxpayers were permitted a deduction for contributions or gifts to charitable organizations "... to an amount not in excess of 15% of ... net income ..." (KixMiller & Baar, 1917, 53). The history of charitable contributions is discussed in Appendix 3C.

The Act was designed to raise additional revenues of over $2.5 billion dollars annually, exclusively for war purposes (Blakely, 1917, 791). The *flat* ordinary or normal tax and the *progressive* additional tax or surtax (previously imposed on net incomes exceeding $20,000, but reduced to include net incomes exceeding $5,000 by the 1917 law) was supplemented by an 8% *war profits and excess profits tax* (KixMiller & Baar, 1917, 1; 1917 Form 1040 Instructions).

The **Revenue Act of 1918**[11] was also referred to as the **War Profits and Excess Profits Tax of 1918**, due to its similarity to tax laws imposed during the Civil War and the emergencies of 1898 and 1914 (KixMiller & Baar, 1919, 5).[12] This Act was effective for the 1918, 1919 and 1920 tax years.

Quarterly tax payments were due (in arrears) for the 1918 tax year. At least 25% of the tax was due, along with the individual's federal income tax return (Form 1040), on March 15th of the next (1919) calendar year. An additional 25% of the tax was due on or before June 15th, September 15th, and December 15th of the next (1919) and future calendar years.

The chief feature of the **Revenue Act of 1921**[13] was to reduce the *surtax* rates (from a maximum of 50% to 40%) component of federal tax, which had reached the maximums established in the **Revenue Act of 1918** (Blakely, 1924, 475). Each of the two preceding fiscal years had closed with surpluses. This Act was effective for the 1921, 1922 and 1923 tax years.

Rebates or refunds of up to 25% of 1923 tax year taxable incomes were provided, and *earned* incomes were taxed at 75% of the rates imposed on other income. This was the first time for differentiation between earned and other income items, and represented the first *earned income tax credit* (EITC).[14] Also, beginning with the 1921 tax year, profit-sharing and pension trusts achieved tax-exempt status.

This Act also introduced a taxpayer "capital net gain" election for assets held for more than two years.[15] This option was available for the 1922 and 1923 tax years. The taxpayer had the option of being taxed at 12.5% on this separable component of income, provided that the election did not result in an overall average federal income tax rate of less than 12.5% (IRS SOI, 1926, 42–3).

Like the **Revenue Act of 1921**, the chief feature of the **Revenue Act of 1924**[16] was to reduce federal income tax rates below the maximums set by the Revenue Act of 1918 (Blakely, 1924, 475). This act first provided for the 12.5% credit for "capital net loss". Previously, such losses were only deductible in arriving at ordinary net income.

The *Revenue Act of 1924* introduced a 25% tax credit for *earned net income* (IRS SOI, 1926, 42–3).[17] It also imposed a tax on donative transfers made by individuals, corporations, associations, partnerships, trusts, or estates. The tax rates imposed on net taxable gifts corresponded to those imposed under the estate tax on the net estate. This was the first time that a gift tax had been imposed by an act of Congress (IRS SOI, 1924, 96).[18]

The *Revenue Act of 1926*[19] applied to individual taxpayers for the 1925–1927 tax years[20] and repealed the gift tax for gifts made after January 1, 1926.

The *Revenue Act of 1928*[21] applied to individual taxpayers for the 1928–1931 tax years. Normal tax rates were reduced by 1% (as amended by the President on December 16, 1929) for the 1929 tax year (IRS SOI, 1928, 450–1), but were restored to their 1928 levels for the 1930 and 1931 tax years (IRS SOI, 1930, 316–7).[22]

The *Revenue Act of 1932*[23] and the *National Industrial Recovery Act*[24] re-imposed a gift tax on gifts made after June 6, 1932. It increased both normal and surtax rates and eliminated the tax credit for earned income. It also limited deductions for losses, to offset gains, from the sale or exchange of stocks and bonds (IRS SOI, 1932, 1).

The *Revenue Act of 1934*,[25] made the personal exemption and exemptions for dependents deductible for the purpose of the *surtax* as well as the *normal* tax. The *Revenue Act of 1935*,[26] and the New Deal's *Social Security Act of 1935*, under a *soak-the-rich* sentiment, increased surtax rates.

The *Revenue Act of 1936*[27] resulted in application of the normal tax to dividends received by domestic corporations and increases in surtax rates (IRS SOI, 1936, 1).

The *Revenue Act of 1938*,[28] revised the treatment of net short-term gains and losses, net long-term gains and losses, and applied an alternative tax for returns with net long-term capital gains and losses (IRS SOI, 1938, 3).[29]

The *Internal Revenue Code of 1939*,[30] in the newly developed section 22(a), provided for the application of the federal income tax to compensation for earned income from personal service (the *Public Salary Tax Act of 1939*) received after December 31, 1938, for government employees and U.S. court judges taking office on or before June 6, 1932.[31] In addition, returns for the 1939 tax year were the first to show the net short-term capital loss carry-over provided by section 117(e) of the IRC (IRS SOI, 1939, 3).

The *Revenue Act of 1940*[32] reduced the amount of gross income for which a return was required (IRS SOI, 1940, 3).

The *Revenue Act of 1941*[33] eliminated the defense tax, imposed increased surtax rates on surtax income, and further reduced the amount of gross income for which a return was required.

The *Revenue Act of 1942*[34] provided for the deductibility of medical and dental expenses and a standard deduction in lieu of itemized deductions, beginning with the 1942 tax year (discussed in Appendix 3D).

Beginning January 1, 1943, the *Current Tax Payment Act of 1943*[35] required the collection of the income tax liability from individuals through withholdings on wages and prepayments on a declaration of estimated tax. This shift to a pay-as-you-go Federal income tax payment system was succinctly characterized by Shlaes (1999, 4):

> ... Ruml, man of ideas ... observed that customers didn't like big bills. They preferred making payments bit by bit ... Ruml devised a plan ... Employers would retain a percentage of taxes from workers every week ... Withholding as we know it today was born.

Initially, withholding was required at a rate of 5%. This amount was increased to 20%, effective for the first complete payroll period following July 1, 1943. Estimated tax payments were facilitated by requiring that persons with large amounts of income not subject to withholding file a declaration of estimated tax for the taxable year on September 15, 1943. These estimated payments were due in two installments: the first on September 15, 1943, and the second on December 15, 1943 (IRS SOI, 1943, 5).

The Second Period of Individual Taxation (1944–1953)

The *Individual Income Tax Act of 1944*[36] made numerous changes affecting individuals. Filing requirements for income tax returns (e.g. gross income limits, normal and surtax exemptions, and special deductions for blindness) were revised and the Form W-2, for withholding on wages, replaced the Form 1040A for taxpayers with total income below $5,000. The victory tax and earned income credit were repealed, normal tax rates were reduced, and surtax rates were increased. Most importantly, deductions for contributions and medical and dental expenses were based on adjusted gross income (AGI) instead of net income (NI) (IRS SOI, 1944, 7–8).

The *Revenue Act of 1948*[37] reduced marginal tax rates, introduced a marital deduction for estate and gift tax purposes, and approved split-income treatment for married couples.

The *Revenue Act of 1950* increased tax rates through the elimination of percentage reductions from *tentative* tax (effective for the 1948 and 1949 tax years) (IRS SOI, 1950, 6). The *Social Security Act Amendments of 1950* (dated August 28, 1950) and the *Revenue Act of 1951*,[38] raised marginal tax rates and imposed the first self-employment tax, levied on the statutory amount of net earnings from self-employment (and including the distributive share of profits or losses from partnerships) at a rate of 2.25%, beginning with the 1951 tax year (IRS SOI, 1951, 6–7).[39]

The Third (and Contemporary) Period of Individual Taxation (1954–)

The *Revenue Act of 1954* (IRC54)[40] made many changes. Tax rates were reduced for widows or widowers, exemptions for children and other dependents and individuals supported by more than one individual were modified, provisions were made for the exclusion of sick pay, medical deductions and contributions rules were liberalized, new dividend provisions were imposed, and retirement tax credits and deductions for child care were made available (IRS SOI, 1954, 101).

The *Technical Amendments Act of 1958*[41] and the *Small Business Tax Revision Act of 1958* resulted in few changes in tax law. One change was the liberalization of the medical deduction for disabled persons age 65 and older, for returns filed for the 1958 tax year (IRS SOI, 1958, 3).

The *Self-Employed Individuals Tax Retirement Act of 1962*[42] treated self-employed taxpayers as employees so that, effective taxable years after December 31, 1962, they could be covered under the same qualified employee retirement plans previously provided for their employees (IRS SOI, 1963, 3).

The *Revenue Act of 1964*[43] introduced major changes for the 1964 tax year, including a reduction of maximum marginal tax rates from 91% to 70% (where they remained through the 1981 tax year). The new legislation provided for higher ceilings for the charitable contribution deduction (discussed in Appendix 3C) and retirement income credits. It also provided greater tax benefits for taxpayers 65 and older on their medical expenses and sales of their personal residences; deductions for moving expenses; restrictions on the sick pay exclusion and the deduction for taxes paid; an increase in the dividend exclusion; a decrease in the dividends received credit; and income-averaging (available for the 1964–1986 tax years and discussed separately in Appendix 3E) (IRS SOI, 1964, 152).

The *Excise Tax Reduction Act of 1965* provided for a systematic reduction or repeal of many of the excise taxes initially levied as emergency revenue-raising measures at the time of the Korean War, World War II, and the depression of the 1930s (CCH, 1965, 3). There were no other basic changes for the 1965 tax year.

For the 1968 tax year, a 10% income tax surcharge was imposed beginning April 1, 1968 (in effect, a 7.5% surcharge for the 1968 tax year). This 10% surtax was extended to the 1969 tax year.

The *Tax Reform Act of 1969* (TRA69)[44] introduced the *maximum tax* (MAXTAX) (discussed in Appendix 3F). TRA69 also modified the treatment of *investment interest* (discussed in Appendix 3G).[45]

For the 1970 tax year, the new tax law introduced a new minimum standard deduction or low income allowance, and a tax on specified tax preferences. Additional changes included the liberalization of the moving expense deduction, limitations on capital loss deductions, and the imposition of higher rates on capital gains (IRS SOI, 1970, iv).

Beginning with the 1972 tax year, a limitation was imposed on the deductibility of *investment interest* (discussed in Appendix 3G).

The *Revenue Act of 1971*[46] provided for a deduction for dependent care expenses.

The *Employee Retirement Income Security Act of 1974* (ERISA74)[47] made major changes to the U.S. private pension system. It (1) imposed additional requirements relating to participation, vesting, and funding for qualified pension and employee benefit plans (for plan years beginning after September 2, 1974), (2) changed the tax treatment of qualified retirement and pension plans by providing for the exclusion of portions of lump-sum distributions (for taxable years beginning after 1973), and (3) increased the retirement contribution deductions available to self-employed taxpayers (for taxable years beginning after 1973). This act also provided for the establishment of "individual retirement accounts" (IRAs), beginning with the 1975 calendar year (IRS SOI, 1974, vi). Initially, the *above-the-line* IRA deduction ceiling, available to employees not covered by qualified retirement plans, was $1,500.

The *Tax Reduction Act of 1975*,[48] generally provided for lower individual federal income taxes, beginning with the 1975 tax year. Prior law[49] had provided for the *above-the-line* deductibility (previously permitted only as an itemized deduction) of penalties paid for the premature withdrawal of funds from savings (IRS SOI, 1974, vi).

The *Tax Reform Act of 1976* (TRA76)[50] implemented numerous tax law changes designed to reduce tax liabilities, including: (1) the replacement of the child care *deduction* with the child care *credit*, (2) a two-year extension[51] (to 1976 and 1977) and expansion of the *earned income credit* (EIC), originally introduced for the 1975 tax year only (discussed in greater detail in Chapter 7), and (3) the replacement of the *retirement income credit* with a more liberal *credit for the elderly* (including phase-outs for high-income taxpayers).

TRA76 also provided for: (1) the introduction of the "at risk" limitation[52] on losses for *tax shelters*, (2) the introduction of two new "tax preference" items, (3) an increase in the *minimum tax*, and (4) the abolition of the sick pay exclusion (IRS SOI, 1976, vi).

The *Tax Reduction and Simplification Act of 1977* established zero-bracket amount exemption deductions.

The *Foreign Earned Income Act of 1978* instituted new provisions for income earned abroad (primarily effective for the 1979 tax year, though transitional provisions were made for the 1978 tax year). The *Revenue Act of 1978* made several changes to the taxation of capital gains (the details are discussed in Chapter 10), tax shelters, employee benefits, and estate and gift taxes, including the one-time exclusion for up to $100,000[53] on gains from the sale of the taxpayer's principal residence, for taxpayers age 55 or over. The *Energy Tax Act of 1978* introduced a residential "energy" credit (for the 1978 tax year) in or on the principal residence of the taxpayer (IRS SOI, 1978, vii).

The *Economic Recovery Tax Act of 1981* (ERTA81)[54] provided for numerous tax law changes to be phased in over several years. It was intended to increase savings and encourage investment, and included: (1) individual federal income tax rate reductions (from a range of 14% to 70% for the 1981 tax year, to a maximum tax rate/bracket of 50% for tax years beginning after 1981), (2) a new *above-the-line* deduction for *two-earner* working married couples (calculated on Schedule W and discussed in Appendix 2H), and (3) the liberalization of rules applying to IRAs, Keoghs, SEPs, and other retirement systems, and providing that all taxpayers, including active participants in qualified plans, could contribute up to $2,000, based on 100% of compensation (IRS SOI, 1982, 9).

ERTA81 provided taxpayers with several tax-saving opportunities, beginning with the 1982 tax year, including: (1) the tax-free generation of up to $2,000 interest on "all savers" certificates, (2) replacement of the $200 ($400 on joint returns) interest and dividend exclusion with a $100 ($200 in joint returns) dividend exclusion, and (3) the introduction of an exclusion from income of up to $750 ($1,500 on joint returns) of public utility dividends (RIA, 1981).

The *Tax Equity and Fiscal Responsibility Act of 1982* (TEFRA82)[55] eliminated the add-on minimum tax, beginning with the 1983 tax year. TEFRA represented the largest revenue-raising bill ever passed, tightening up on itemized deductions and pension rules. Additional changes affecting individual taxpayers included the reduction of the *base amount* for the taxation of unemployment compensation and increases in both medical deduction (raised to 5% of AGI) and casualty deduction (raised to 10% of AGI) floors (RIA, 1982).

From early January–mid-August 1983, tax laws were either directly or indirectly influenced by the *Orphan Drug Act*,[56] the *Highway Revenue Act*,[57] the *Technical Corrections Act*,[58] the *Periodic Payment Settlement Act*,[59] the *Interest and Dividends Tax Compliance Act* and *Caribbean Basin Economic Recovery Act*,[60] and the *Railroad Retirement Benefit Act*.[61] Beginning with the 1984 tax year, individual taxpayers were required to include in gross income part of the social security, railroad retirement, and sick pay benefits received. The disability income exclusion was eliminated and rules for the elderly credit were changed (RIA, 1983).

The *Social Security Amendments of 1983*[62] raised payroll tax rates and wage bases to avoid the impending shortfall and avoid benefits reductions for retirees. The history of the U.S. Social Security system is described in greater detail in Chapter 11 and related Figures and Tables.

The *Tax Reform Act of 1984* (TRA84)[63] and the *Spending Reduction Act of 1984* together comprised the *Deficit Reduction Act of 1984* (DEFRA84), which included the imposition of imputed interest requirements on *below-market*-interest loans, reduced the holding period for long-term capital gains, slashed income-averaging provisions, and provided statutory exclusions for employee fringe benefits.

The *Consolidated Omnibus Budget Reconciliation Act of 1985* (COBRA85) affected capital gains and alternative minimum tax rules. The *Tax Reform Act of 1986* (TRA86)[64] resulted in the most significant and complex tax revision in U.S. history, re-designating the tax law as the *Internal Revenue Code of 1986* (IRC86). It restricted deductions for personal interest, imposed restrictions on passive activity losses (discussed in Appendix 3I), IRA contributions, miscellaneous itemized deductions, and deductions for state and local sales taxes.

The *Revenue Act of 1987*[65] made few notable changes. The *Pension Protection Act of 1987*[66] affected funding requirements for qualified pension plans (RIA, 1988).

The *Family Support Act of 1988* reduced the tax benefits associated with children and dependent care, for tax years beginning: (1) after 1988 and (2) 1989. For tax years beginning in 1989, additional amounts were subject to itemizing as miscellaneous deductions. The *Technical and Miscellaneous Revenue Act of 1988* (TAMRA88)[67] made corrections and technical adjustments to TRA86 and included a taxpayer's bill of rights.

The *Revenue Reconciliation Act of 1989* (RRA89)[68] achieved deficit reduction by reducing the benefits associated with employee stock ownership plans (ESOPs),[69] and imposing slight increases on and accelerated the collection of withholding and payroll taxes.

The *Omnibus Budget Reconciliation Act of 1990* (OBRA90)[70] introduced personal exemption phase-outs, increased from two to three the number of statutory rates, and provided for a maximum capital gain rate of 28%.

The *Tax Extension Act of 1991* extended, for six months, 11 tax provisions scheduled to expire on December 31, 1991; the *Energy Policy Act of 1992* increased the exclusions available for employer-provided transportation benefits; and the *Revenue Reconciliation Act of 1993* (RRA93)[71] raised tax rates for moderate- and high-income taxpayers and modified the laws affecting the AMT and passive losses.

During 1994, tax legislation included the *Social Security Domestic Employment Reform Act of 1994*, primarily affecting payroll and unemployment taxes paid for domestic workers, and the *General Agreement on Tariffs and Trade* (GATT), involving revenue raising measures to offset tariff reductions.

During 1996, tax legislation included the *Taxpayer Bill of Rights 2*,[72] the *Small Business Job Protection Act*,[73] the *Health Insurance Portability and Accountability Act*,[74] and the *Personal Responsibility and Work Opportunity Reconciliation Act*,[75] primarily affecting the EITC.

The *Taxpayer Relief Act of 1997*[76] provided tax cuts, including reduced capital gains tax rates, and resulted in the expansion of IRAs, a child tax credit, tax incentives targeted toward education, and the reduction of estate taxes.

Tax legislation, during 1998, included the *IRS Restructuring and Reform Act*,[77] the *Transportation Equity Act for the 21st Century*,[78] and the *Tax and Trade Relief Extension Act*.[79]

The *Tax Relief Extension Act of 1999* (HR 1180) received final Congressional approval on November 19, 1999.

SUMMARY

This chapter has provided a broad overview of the history of U.S. individual federal income taxation. Appendix Tables 3-1 and 3-2 provide summaries of both U.S. and Canadian descriptive statistics respectively, including the number of individual income tax returns filed. These descriptive statistics were used to generate Fig. 3-2.

Appendices 3C–3I are devoted to selected topics, only briefly addressed in this chapter, that

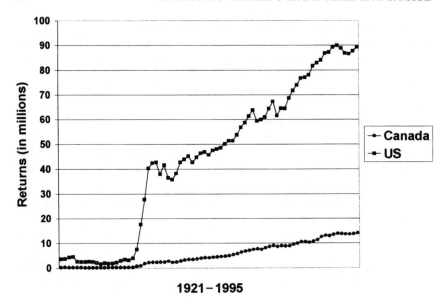

Fig. 3-2. A Comparison of U.S./Canadian *Taxable* Individual Federal Income Tax
Returns (1921–1995).

N = 75
r = 0.955 (Pearson correlation coefficient; p-value = 0.000)

(1) Have been topics of academic research and policy discussions, or
(2) Represent policy solutions to past problems or inequities, but
(3) Are not likely to be of great interest in the *immediate* future.

For example, income-averaging (1964–1981) and the maximum tax
(1971–1981), discussed in Appendix 3E and 3F, respectively, were designed to
minimize inequities evolving from steeply progressive tax rates (e.g. 14% to
70%). They pre-dated a move to the contemporary (relatively flatter)
progressive tax rates (e.g. 15% to 39.6%). Should the U.S. system ever return
to a steeply progressive tax rate structure, these techniques may again resurface
to represent policy solutions.

The topics addressed in Appendices 3C–3I are summarized in Table 3-1.

Table 3-1. Summary of Topics Covered in Appendices (App.) to Chapter 3.

App.	Topic	Current Status	Policy Rationale
3C	Charitable Contributions	Itemized Deduction; <50% AGI	Subsidizes substitutes for government programs
3D	Medical and Dental Expenses	Itemized Deduction; >7.5% AGI	Pursues equity with subsidy for families with reduced ability to pay
3E	Income-Averaging (1964–1981)	**EXPIRED**	Pursues equity under steeply progressive tax rates schedules
3F	Maximum Tax (MAXTAX) (1971–1981)	**EXPIRED**	Pursues equity under steeply progressive tax rates schedules
3G	Investment Interest	Itemized Deduction; Limitations	Pursues equity with limitation for high income taxpayers
	Personal Interest (1913–1990)	**EXPIRED**	Reduced economic incentives for personal consumer debt
	Mortgage Interest	Itemized Deduction; Limited to $1.1M	Pursues equity with limitation for high income taxpayers
3H	Two-Earner Married Couple Deduction (1982–1986)	**EXPIRED**	Intended to reduce economic hardships on married taxpayers (see Chapter 8)
3I	Passive Activity Losses (PALs)	Itemized Deduction; Limitations	Pursues equity with limitation for high income taxpayers

CHAPTER 4
PROGRESSIVE TAXATION AND ALTERNATIVES

The question of tax progressivity – who should bear the tax burden – has fascinated tax philosophers for over a century, and remains highly controversial . . . (Slemrod, 1994, vii).

The modern approach to evaluating progressivity focuses on the trade-off between the potential social benefit of a more equal distribution of income and the economic costs caused by the disincentive effects of the high marginal tax rates required by a distributive tax system (Slemrod, 1994, 2–3)

INTRODUCTION

Slemrod (1994, 2) notes that economists have traditionally used two principles to evaluate how tax burdens *should* be allocated: (1) the benefits principle and (2) the ability-to-pay principle. The benefits principle sees taxes as a quid pro quo for government services provided and tax burden is applied based on the benefits received by each citizen. However, the benefits principle has operational problems (e.g. how does each individual benefit from national defense?) and excludes the possibility of governmental income redistribution. The ability-to-pay principle suggests that progressive taxation leads to a sacrifice-based form of income equalization, but does not answer the question as to how fast tax rates should rise.

This chapter provides a brief examination of the principles upon which the American tax system's focus and application of progressive tax rates is founded. The alternatives to progressive income taxation, namely, a national sales tax, a flat or proportional tax, and a value-added tax (VAT), are also discussed.

Adam Smith's four desirable principles of certainty, equality, convenience, and economy continue to represent the primary focus of the U.S. system of individual income taxation. Strayer (1939) employed a comparable framework in his examination of *small* incomes, as the U.S. system moved from a *class* tax to a *mass* tax (see Appendix 4J). Even the Canadian tax system (1917–), which applied U.S. tax principles (1913–) through the post-WWII period, has retained many of the basic principles of U.S. taxation (see Appendix 2B).

An exception is represented by the Canadian trend toward the use of personal exemption tax credits. This trend has been adopted for use in the American system of taxation, but only on a piecemeal or supplemental basis (discussed in Chapter 6).

It is not uncommon for U.S. policy makers to examine Canadian systems (Shoven & Whalley, 1992). For example, the Canadian health care system has been and continues (though to a lesser degree) to be examined as a possible replacement for the American healthcare system. Contemporary *quality of care* issues, representing failures of the Canadian health care system, have led to a general failure to support a U.S. conversion to the Canadian system.

Similarly, the Canadian tax system has led the U.S. in a move toward provincial and national sales taxation. It is the failure of this Canadian sales tax system to represent a vehicle for *replacement* of the Canadian individual income tax that has hindered its adoption in the U.S. Though study, simulations, debate and policy discussion is likely to continue, we do not anticipate the implementation of a U.S. national sales tax in the immediate future. This raises questions as to how the U.S. will deal with the issue of Internet sales taxation (see Appendix 4K).

BASIC HISTORICAL AND CONTEMPORARY U.S. PRINCIPLES OF TAXATION

An indirect tax may be called a tax on commodities that is paid by the purchaser. The objection to indirect taxation is that being a tax on consumption it does not discriminate between the poor man's last dollar and the income of the millionaire. Another objection . . . is the cost required in its collection (Burke, 1891, 252–253).

Strayer (1939, 5) noted the "substantial uniformity" in the broad conception of the fundamental principles of taxation. He recalled that Adam Smith, in *The Wealth of Nations* (Book V, Ch. II, Part ii), specified the desirable principles of: (1) certainty, (2) equality, (3) convenience, and (4) economy.

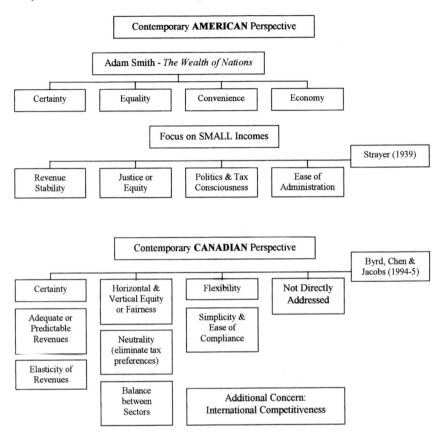

Fig. 4-1. Basic Principles of Taxation in the U.S. and Canada.

Strayer (1939) decided to structure his examination of the taxation of *small* incomes around the following, comparable framework:

(1) Revenue stability or self-adjustment to the changing needs of government,
(2) Social and ethical considerations or justice and equity issues (discussed in Chapter 5),
(3) Politics or the level of tax consciousness, and
(4) Ease of administration.

Figure 4-1 illustrates how the American tax system fits into Adam Smith's conceptual framework. This figure also provides a comparison with the Canadian system.

BASIC CONTEMPORARY CANADIAN PRINCIPLES OF TAXATION

> While progressive tax systems continue to be pervasive, there has been a worldwide trend towards flattening rate schedules. One of the reasons for this trend is the fact that effective tax rates are not as progressive as the rate schedules indicate ... high bracket taxpayers tend to have better access to various types of tax concessions, a fact which can significantly reduce the effective rates for these individuals. Given this situation, it has been suggested that we could achieve results similar to those ... by applying a flat rate of tax to a broadened taxation base (Byrd, Chen & Jacobs, 1994–95, 5).

Byrd, Chen and Jacobs (1994–1995), authors of a Canadian taxation textbook, summarize what they believe are the desirable qualitative characteristics of a tax system (6–7). Their list (see Fig. 4-1) is comparable to those contained in U.S. texts, and includes:

- Equity Or Fairness – Horizontal and vertical equity (see Chapter 5).
- Neutrality – The elimination of tax preferences or statutorily imposed taxpayer bias.
- Adequacy – A system that produces sufficient and predictable tax revenues.
- Elasticity – A system which results in changes in tax revenues without the need to impose tax rate changes.
- Flexibility – The degree of ease in tax system adjustment to meet changing socioeconomic needs.
- Simplicity and Ease Of Compliance – A system that taxpayers and enforcers of the tax system find easy to understand and comply with.
- Certainty – Tax payment amounts, the basis of tax payments, and the tax payment dates should be clearly determinable to assist taxpayers and taxing authorities in budgeting and forecasting.
- Balance Between Sectors – Significant reliance on any particular sector (e.g. individual or corporate) should be avoided, as should the imposition of a disproportionate portion of the tax burden on any single sector.
- International Competitiveness – Individuals and businesses will gravitate (to the extent practicable) toward those countries with relatively favorable tax rates.

Byrd, Chen and Jacobs (1994–1995, 5) argue that the Canadian introduction of the goods and services tax (GST, beginning with the 1991 tax year) may represent the beginning of a change in the Canadian approach to taxation and a shift away from the heavy reliance on progressive rates used for individual and corporate income taxation. They note that progressive tax rates are associated with the following problems:

- Complexity – Efforts by taxpayers to maximize use of the lower tax brackets, in a progressive tax system, require the development and use of complex anti-avoidance rules by taxation authorities.
- Income Fluctuations – Progressive tax rates discriminate against individuals with highly variable taxable incomes.
- Family Unit Problems – Progressive tax rates discriminate against single income family units.
- Economic Growth – High progressive tax rates can discourage both employment and investment efforts, limit economic growth, and even result in lower aggregate tax collections.
- Tax Concessions – Taxpayers in high progressive tax brackets are provided with an economic incentive to devote significant economic resources to seek the development of favorable tax provisions that they might take advantage of.
- Tax Evasion – Progressive tax rates discourage the reporting of all income and encourage the creation of tax evasion mechanisms and/or criminal activities.

Both U.S. and Canadian individual federal income tax systems remain progressive, though flatter than historical rates. This element of *complexity*, the first principle discussed by Byrd, Chen and Jacobs, does not appear to merit contemporary concern. Presently, the U.S. has five progressive tax rates or brackets and the Canadian system has three. However, the complexity of the U.S. or Canadian tax systems is not a function of progressive (as opposed to flat) tax rates, but one of laws affecting different categories or classifications of income.

The second element discussed by Byrd, Chen and Jacobs is that of *income fluctuations*. The primary remedy used in the U.S. was income-averaging. As progressivity declined, the need in the U.S. for income-averaging (discussed in Appendix 2E), as well as the MAXTAX (discussed in Appendix 3F) was lessened and these measures were repealed.

The third element, *family unit problems*, has been characterized in the U.S. as the marriage tax penalty (MTP). In Canada, the *individual* is the unit of taxation. In the U.S., the *household* is the unit of taxation. Though this topic was introduced in Appendix 3H and is addressed in greater detail in Chapter 8, it is relevant to note that, despite the complete absence of a MTP in Canada, divorce rates (as a percentage of marriages) are approximately the same in the U.S. and Canada.

The fourth principle discussed by Byrd, Chen and Jacobs is that of *economic growth*. They suggest that high progressive tax rates, a condition that no longer exists in either the U.S. or Canada, may inhibit economic growth – even to the extent of lower tax collections.

The fifth issue is one of *tax concessions*. The argument is that high-income taxpayers are provided with an economic incentive to fund the pursuit of favorable tax concessions. However, ceilings on home mortgage and invest-ment interest (discussed in Appendix 3G), as well as the implementation of limitations on passive activity losses (discussed in Appendix 3I) adversely affected high-income taxpayers, and would appear to provide evidence to the contrary, at least with respect to the *complete* success of such efforts.

Finally, Byrd, Chen and Jacobs suggest that progressive tax rates discourage the reporting of income and encourage *tax evasion*. This may be true. However, it is unlikely that tax evasion would disappear under a completely flat income tax system.

In a discussion inclusive of the Canadian general sales tax (GST), Byrd, Chen and Jacobs (1994–1995, 7–8) concede that the Canadian system of taxation remains: (1) somewhat *inequitable* (despite their own counterpart to the U.S. alternative minimum tax (AMT), designed to partially alleviate some of the above inequities), (2) relies heavily on *personal* income taxes for total revenues, (3) is plagued by problems with *stability* and *dependability* (though the GST is intended to at least partially correct for this weakness), (4) is very *complex* (worsened by the GST), and (5) "(w)ithout question", is lacking in the desirable *international competitiveness* feature, particularly true when com-pared with the U.S., as Canada has experienced a significant loss in business activity and skilled workers.

Byrd, Chen and Jacobs (1997–1998, 62) suggest that the Canadian implementation of the GST (1991–) contributes positively to the Canadian tax system, as follows: (1) *simplicity*, (2) *consistency*, (3) *international com-petitiveness* (by lowering Canadian marginal individual and corporate tax rates), and, for similar reasons, (4) results in the maintenance of economic *incentives to work*.

THE CANADIAN NATIONAL SALES TAX

. . . the sales tax is regarded as the easy way to collect considerable revenue from the lower strata of income recipients and the income tax is the means of insuring that the rich pay "their share" (Groves, 1963, 21).

One of the major problems with the goods and services tax (GST) introduced in Canada in 1991 is the fact that it is a regressive tax. In order to provide some relief from the impact of this characteristic on low income families, accompanying legislation provided for a refundable GST credit under ITA 122.5 (Byrd, Chen & Jacobs, 1997–1998, 408).

The Canadian GST replaced Canada's manufacturers' sales tax (MST) on January 1, 1991 (Treff & Cook, 1995, 4: 3). The GST is a form of value-added tax (VAT). Initially, the GST was 7%.

All purchasers pay GST. Sellers must register with Revenue Canada, the Canadian counterpart to the United States' Internal Revenue Service (IRS). The Canadian government uses the GST as a mechanism for the preferential treatment of selected purchases.

Intended to be a tax on final consumption only, the Canadian GST is imposed on the sales of all goods and services at all stages of production and consumption. Businesses are provided with tax credits for refunds of all GST paid on *cost of goods sold* or *inputs* to production. Tax credits (or refunds) are not available to the ultimate consumer, though a refundable tax credit is provided for on the Canadian FIT return.[80]

Nine of the ten Canadian provinces imposed a provincial (retail) sales tax (PST) in addition to the federal GST. (Appendix Figure 4-1 provides a labeled map of the Canadian provinces.) Neither of the two Canadian territories were subjected to a PST. For 1995, PST and combined GST (7%) and PST rates for all Canadian provinces and territories were imposed at the following rates (Treff & Cook, 1995, 5: 3–5: 4):

(1)	Newfoundland	$12.0\% + 7\% = 19\%$,
(2)	Prince Edward Island (PEI)	$10.0\% + 7\% = 17\%$,
(3)	Nova Scotia (NS)	$11.0\% + 7\% = 18\%$,
(4)	New Brunswick (NB)	$11.0\% + 7\% = 18\%$,
(5)	Quebec	$6.5\% + 7\% = 13.5\%$,
(6)	Ontario	$8.0\% + 7\% = 15\%$,
(7)	Manitoba	$7.0\% + 7\% = 14\%$,
(8)	Saskatchewan	$9.0\% + 7\% = 16\%$,
(9)	British Columbia (BC)	$7.0\% + 7\% = 14\%$,
(10)	Alberta	$-0\% + 7\% = 7\%$,
(11)	Northwest Territory	$-0\% + 7\% = 7\%$, and
(12)	Yukon Territory	$-0\% + 7\% = 7\%$.

In addition to the above, many provinces provide for separate sales taxes on alcoholic beverages, restaurant meals, telephone services, etc. This aspect of the Canadian system is comparable to that of the United States, where excise taxes are imposed on alcoholic beverages, telephone services, etc.

U.S. PROPOSALS FOR NATIONAL SALES TAXATION

. . . we find that the tax burden on high income households is generally lower under the retail sales tax than under the income tax. This pattern arises from the greater share of income that is devoted to consumption at low income levels (Feenberg, Mitrusi & Poterba, 1997, 44).

. . . the required tax rate in a national retail sales tax (NRST). . . would be over 50% . . . (Gale, 1999, 443).

. . . there are problems with a national sales tax. One is this: in nearly every country where it has been tried, whatever the political promises, a sales tax has come *in addition* to an income tax, not instead of one (Shlaes, 1999, 224).

Recently, proposals to replace most federal taxes with a national retail sales tax. . . include H. R. 2001. . . and the "Fair Tax" plan. . . (Koenig, 1999, 683).

Hubbard (1996, 1) suggests that those interested in restructuring the U.S. tax system are motivated by a desire to "simplify compliance and improve incentives". He suggests that replacements for the federal (corporate and individual) income tax include four *consumption-based reforms*:

(1) A retail sales tax: a proportional tax paid by businesses, for consistency,
(2) A "flat" tax: similar to a value-added tax (VAT) for business; little change to the existing personal tax, but taxing only wages/salaries, which are deducted from the base of the business tax,
(3) The Unlimited Savings Allowance (USA) tax:[81] similar to a VAT for business; little change to the existing personal tax, but taxing only "consumed income", without deductions for wages/salaries from the business, but with tax credits for payroll taxes, and
(4) A VAT.

Hubbard suggests that the present federal income tax fails to treat investment uniformly and is affected by inflation. Alternatively, he suggests that a consumption-based tax treats different kinds of investments uniformly and is inherently inflation proof. He also addresses the transitional issues of a one-time tax on "old savings" or "old capital". International and corporate issues, the topical coverage of which is beyond the scope of this monograph, are covered in the same series (Grubert & Newlon, 1997; Gentry & Hubbard, 1997).

Feenberg, Mitrusi and Poterba (1997, 30) examined four possible retail sales tax scenarios:

(1) One without any exemptions,
(2) One with exemptions for food,

(3) One with exemptions for food, medical services, and housing, and
(4) One coupled with a "demogrant" system and providing cash assistance to reduce the tax burden on low-income households.

They considered retail sales taxes that

• Replace only the federal income tax, and
• Replace both the federal income tax and payroll taxes.

Koenig (1999, 683) examined "HR 2001, sponsored by Congressmen Dan Schaefer and Billy Tauzin, and the "Fair Tax" plan put forward by the group Americans for Fair Taxation (AFT). "Both proposals would eliminate the federal income tax and federal gift and inheritance taxes. The AFT proposal would also eliminate federal payroll taxes". Koenig (695) concluded that the correctly calculated program-neutral sales tax rate would exceed 60% (equivalent to a 37% average marginal rate – a rate higher than that facing most U.S. households under current law), if intended to replace both federal income taxes and federal payroll taxes, and 35% if not intended to replace payroll taxes. Furthermore, he suggested that the establishment of a national retail sales tax: (1) would not have lasting positive labor supply effects, (2) would probably result in real stock price declines, and (3) may create some very difficult enforcement problems.

We do not believe that a national retail sales tax (NRST) is either desirable or politically feasible in the early decades of the 21st century. Like the Canadian health care system, those debating this topic are likely to point to the failures of the Canadian GST and PST to *replace* the Canadian income tax.

THE FLAT OR PROPORTIONAL TAX

A "flat" or "proportional" tax requires the use of the same statutory tax rate for all taxpayers at all income levels. In its pure form, a flat tax does not provide for income exclusions, deductions, or exemptions, though contemporary proposals (e.g. the Armey-Shelby-Craig[82] and Specter[83] proposals of the mid-1990s) involve the use of exemptions. According to research conducted by Iyer, Seetharaman and Englebrecht (1996, 83):

> Both (Armey-Shelby-Craig and Specter) flat tax proposals moderate before-tax income inequality modestly, but neither moderates income inequality as effectively as the current income tax system.

We do not believe that there is a high probability that some modified form of flat or proportional tax will be imposed, replacing the present progressive system of Federal income taxation for individuals, in the early decades of the 21st century.

A VALUE-ADDED TAX

Value Added Tax. A tax levied on the increase in value of a commodity that has been created by the taxpayer's stage of the production or distribution cycle (Byrd, Chen & Jacobs, 1994–95, 1).
   . . . proponents of the credit method VAT are correct that it is superior to a standard RST, although some of their arguments are somewhat exaggerated. Passage of a VAT may, however, be politically infeasible in the U.S. (Zodrow, 1999, 429).

The value-added tax (VAT) is used in European countries. Advocates suggest that the VAT is the most efficient method of raising tax revenues. Opponents argue that it is comparable to a national sales tax and is unfair, due to its regressive nature. Because it is a consumption tax, the greatest burden falls on those required to consume all of their income and least able to afford it. Alternatively, those able to save and/or invest a large portion of their income are the major beneficiaries.

   Generally, the VAT requires that each business, at each stage of production, claim a credit for the VAT paid for its *cost of goods sold*. It remits to the appropriate governmental entity the VAT collected on its *sales* or *outputs*, claiming a credit for the VAT paid on *cost of goods sold* or *inputs*. This process is repeated at every stage of production, though the ultimate consumer bears the entire cost of the VAT.

   We concur with Zodrow (1999) and believe that the VAT is an unlikely and even politically infeasible choice for the U.S. for at least the first few decades of the new millenium.

SUMMARY

Progressive taxation is likely to remain the system of choice in the U.S. for the early decades of the 21st century. From the perspective of the lower- and middle-classes, progressive taxation is the most *equitable* system of taxation. Progressive taxation is consistent with the *ability-to-pay* principle of taxation.

   Over the past two decades, researchers have invested considerable effort in the study of replacement of the U.S. progressive individual income tax with a national retail sales tax (NRST). Both the NRST and VAT are flat, consumption-based taxes, regressive in their affect on the population, and unlikely to serve as adequate *replacements* to the contemporary U.S. individual income tax. We do not anticipate passage of a NRST or a VAT in the early decades of the 21st century.

   A flat or proportional income tax represents a politically palatable, and far more likely, replacement to the current system of U.S. individual income taxation. Though less equitable, from the perspective of the lower- or middle-class taxpayer's perspective (e.g. regressive), a flat or proportional tax is likely to be *perceived* as "fair" by many U.S. citizens.

A contemporary issue in U.S. taxation is "how to tax the Internet". Because the vast majority of the sales taking place on the Internet are "business-to-business" (B-to-B), and because B-to-B sales in the U.S. are generally exempt from sales taxation, this topic is likely to be studied further before legislation is passed. Appendix 4K summarizes some of the contemporary events associated with the suspension of federal legislation on Internet sales taxation.

CHAPTER 5
EQUITY OR FAIRNESS ISSUES

... practically all tax authorities agree that justice or equity is an important goal, yet few agree as to the specific measures to be used in seeking justice (Strayer, 1939, 4).

... the use of measures of income inequality is dismissed for two reasons. First, more information is desired about the distribution of income than can be condensed into a single inequality measure. A general index of inequality describes the shape of the complete distribution only under very specialized assumptions. Second, a single measure of inequality, as traditionally defined, may provide an inadequate or misleading picture to person's interested in income inequality itself (Metcalf, 1972, 9).

... there is virtual unanimity that horizontal equity – the extent to which equals are treated equally – is a worthy goal of any tax system ... the measurement of vertical inequality ... is relatively straightforward ... (Auerbach & Hassett, 1999, 1).

The concept of horizontal equity is based on the idea that there are classes of individuals whom we would wish to label "equals" (Auerbach & Hassett, 1999, 4).

INTRODUCTION

Chapter 4 introduced the concepts of: (1) the *benefit principle*, and (2) the *ability-to-pay principle*. The former suggests that those who benefit from the public expenditure should pay the tax that finances it. The latter suggests that those better able to pay should bear the greater burden of taxes, whether or not they benefit more. The ability-to-pay system presumes the presence of both (a) *horizontal equity*, and (b) *vertical equity* (Ruffin & Gregory, 1993, 823–824)

This chapter provides a primer on horizontal and vertical equity. The concepts of explicit and implicit taxes are also discussed. These concepts and quantitative techniques are frequently used by academics in their evaluation of the relative fairness or equity of alternative tax proposals.

Appendix 5L provides a brief discussion of both U.S. and Canadian solutions to "tax bracket creep". The mechanisms used by these countries differ and provide insights into efforts to correct what many American believed to be an "unfair" or "inequitable" condition exacerbated by the double-digit inflation rates occurring in the late 1970s and early 1980s.

HORIZONTAL AND VERTICAL EQUITY

Horizontal equity exists when those with equal abilities to pay do, in fact, pay the same amount of tax. This tax policy concept suggests that taxpayers in similar economic circumstances and of like ability to pay taxes bear similar tax burdens. For example, tax burdens should not differ between taxpayers receiving salary income and those receiving profits from exercising stock options or realizing long-term capital gains.

The latter condition of *vertical equity* exists when taxpayers with a greater ability to pay bear a heavier tax burden. This tax policy concept also suggests that the taxpayer's tax burden should be a function of *ability to pay*. Taxpayers with higher incomes should pay higher taxes (e.g. flat or progressive taxation). This notion is founded on the assumption that the dollar support levels required to provide basic needs and sustain taxpayers might be presumed comparable. Alternatively stated, vertical equity ". . . assumes that, in terms of utility, a dollar loss in income is worth less to a high-income taxpayer than to a low-income taxpayer" (Hyman, 1987, 313).

The calculation of horizontal equity and vertical equity is complicated by the introduction of the concept of *implicit taxes*, which complements the more commonly investigated and easily calculated *explicit taxes*. Figure 5-1 represents a graphic framework for these terms and concepts. The development of the concept of implicit *taxes*, a specific tax-based application of the concept of implicit *costs*, is credited to the research efforts of Scholes and Wolfson (1992).

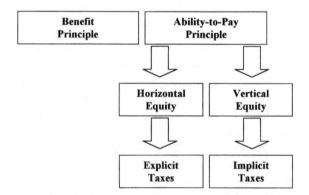

Fig. 5-1. Fairness and Equity Principles.

METHODS USED TO MEASURE EQUITY

If the size distribution of personal income is to be analyzed . . . a convenient mechanism must be found to describe the size distribution. Traditionally, the important feature of personal income has been the degree of income inequality . . . attention has been directed specifically to the lower tail of the income distribution; measures of income inequality have been replaced by measures of the portion of the cumulative distribution lying below some absolute income level or conforming to some definition of poverty (Metcalf, 1972, 8).

Both (horizontal equity and vertical equity) are subjective . . . Insofar as individual assessments of economic capacity differ, there will be differing views concerning horizontal equity. Vertical equity requires judgements on income distribution that are even more subjective . . . (Hyman, 1987, 313).

Metcalf (1972) examined the post-WWII period – the mid-1960s (1947–1965). He selected "the displaced lognormal distribution, a normal transformation corresponding formally to the three-parameter lognormal distribution" (1972, 9) for his study of the size distribution of income. His objective was to develop an alternative to existing econometric models of income distribution.

He identified the Gini concentration ratio or Lorenz measure as the most commonly used measure of income inequality (1972, 9). He also discussed "direct estimation of the population in absolute income cells" (1972, 11), the "characterization of the distribution by a functional form" (1972, 12), and use of the "lognormal distribution" (1972, 12), which competes with the Pareto-Lévy law (1972, 14). Metcalf also discusses the Pearson distribution and normal transformations, displaced lognormal distributions (including parameter estimation), and other normal distributions (1972, 18–25). This and alternative methods are discussed below.

EXPLICIT AND IMPLICIT TAXES

. . . **explicit costs** . . . are the costs that the accountant will be able to measure . . . the **implicit cost** . . . (represents) **(o)pportunity costs** (emphasis added) incurred . . . not explicitly recognized (e.g. foregone rent on land owned and used by a firm) (Miller, 1978, 189 and 216).

. . . investors pay taxes **explicitly** on heavily taxed investments and they pay taxes **implicitly** on lightly taxed investments through lower before-tax rates of return (Scholes & Wolfson, 1992, 84).

Explicit taxes are easily identified. For example, an itemizing homeowner with a mortgage receives a tax benefit, in the form of the home mortgage interest deduction. The tenant or non-homeowner/renter does not. The homeowner enjoys a tax benefit not available to the non-homeowner, because home ownership is a tax-favored investment.

Implicit taxes suggest the presence of an efficient market that "bids up" the before-tax cost of tax-favored investments (e.g. home ownership). Therefore, the price of home ownership would, in an efficient housing market, be expected to rise (decline) with an increase (decline) in the tax benefits or savings associated with home ownership.

Scholes and Wolfson (1992, 87) defined the *implicit tax rate* as follows:

$$R_a = R_b(1 - t_{la}) \tag{5-1a}$$

or

$$t_{la} = (R_b - R_a) \div R_b \tag{5-1b}$$

where t_{la} represents the implicit tax rate on a particular investment (a), R_b denotes the risk-adjusted before-tax return on fully taxable bonds, and R_a denotes the risk-adjusted before-tax return on some alternative investment.

Scholes and Wolfson (1992, 85) provided several examples of *tax-favored* and *tax-disfavored* investments in the United States and other countries, as follows:

Tax-favored investments include:

- The full tax exemption for U.S. municipal bonds,
- The partial exemption for capital assets in most countries,
- Tax credits (e.g. the investment tax credit, targeted jobs credit, alcohol fuel credit, research and experimentation credit, low-income housing credit, energy investment credit, payroll tax credit, and the rehabilitation investment credit in the U.S.),
- Accelerated tax deductions (e.g. immediate expensing of research and experimentation costs, advertising expenditures, personnel costs incurred to expand into new markets, and accelerated depreciation), and
- The failure to tax appreciation in the value of many assets.

Tax-disfavored investments include:

- Items subjected to special or additional tax assessments (e.g. the Windfall Profits Tax on oil, import duties, and excise taxes),
- Accelerated taxable income (e.g. risky bonds with high coupon rates that include a default premium equivalent to a return of capital), and
- Decelerated tax deductions (e.g. pre-Internal Revenue Code Section 197 non-amortizable goodwill or trademarks).

Favorable tax treatment (such as liberal depreciation allowances or investment tax credits) stimulates demand for investments. Increased investment exerts upward pressure on factor prices (e.g. labor costs and equipment costs), unless the supply of such factors is perfectly elastic, and downward pressure on consumer prices unless consumer demand is perfectly elastic (Scholes & Wolfson, 1992, 95).

Generally, we discuss U.S. individual federal income tax issues from the *explicit* tax perspective. We do not attempt to calculate *implicit* taxes.

METHODS USED TO MEASURE HORIZONTAL INEQUITY

Iyer and Seetharaman (1999) examined the strengths and limitations of three popular contemporary methods[84] for measuring horizontal inequity:

(1) The coefficient of variation (CV) method,[85]
(2) The rank preservation (RP) method,
(3) The decomposition (DE) method.

Many studies of horizontal equity have also examined *explicit taxes* (Madeo & Madeo, 1981; Anderson, 1985, 1988; Pierce, 1989; Ricketts, 1990; Grasso & Frischmann, 1992; Luttman & Spindle, 1994).

(1) The Coefficient of Variation Method

The coefficient of variation (CV) is a measure of relative variation (Ingram & Monks, 1992, 46). It expresses the standard deviation as a percentage of the arithmetic mean. Therefore, the CV is independent of the units of measurement, as follows:

$$CV_n = (\sigma_n \div \mu_n) \times 100 \qquad (5\text{-}2)$$

where CV = the coefficient of variation of effective tax rates or the taxpayer's tax liability divided by taxable income before-tax for group n, σ = the standard deviation of μ for group n, and μ = the mean effective tax rate for group n. The tax system or proposal generating the lowest CV provides the greatest horizontal equity within a given taxpayer group (n) of "equals".

(2) The Rank Preservation Method

The rank preservation (RP) method measures horizontal equity as a function of pre-tax versus post-tax rerankings of taxpayers. The Atkinson-Plotnick index (API) and the Rank Divergence index (RDI) (Iyer & Seetharaman, 1999) provide two examples of modifications of a Lorenz curve-based approach to the measurement of horizontal equity.

The API is broadly summarized, as follows:

$$API = (\Sigma |M^* - M|) \div (2\Sigma |M - M'|) \qquad (5\text{-}3a)$$

where $M^* =$ a rank-preserving or horizontally equitable level of after-tax income, $M =$ the actual level of after-tax income, and $M' =$ the actual level of after-tax income.

The RDI is broadly summarized, as follows:

$$RDI = (\Sigma | R^* - R |^a)^{1/a} \div MAX \qquad (5\text{-}3b)$$

where $R^* =$ a horizontally equitable before-tax rank, $R =$ the actual after-tax rank, $a =$ a measure subjectively determined by the researcher to capture societal aversion to horizontal inequity, and $MAX =$ the maximum possible measure for the expression in the numerator.

(3) The Decomposition Method

The decomposition method mathematically partitions changes in income inequality into: (1) vertical, (2) horizontal, and (3) reranking components, requiring a reconciliation or distinction between before-tax and three separable after-tax Gini coefficients (Iyer & Seetharaman, 1999).

The Gini coefficient provides a measure of equality or inequality in a distribution. It is the arithmetic mean of all possible differences among the observations in a distribution, regardless of sign, divided by twice the mean of the distribution. The Gini coefficient equals the ratio of the area between the Lorenz curve and the 45-degree line (i.e. a line of equal distribution) to the entire area below the 45-degree line. Its values range from 0 (zero), indicating perfect equality, to 1 (one), indicating perfect inequality.

The Gini coefficient (G) is broadly summarized, as follows:

$$G = (\Sigma\Sigma' | y - y' |) \div (2N^2 \mu) \qquad (5\text{-}4)$$

According to Metcalf (1972, 9–10), "(t)he Lorenz measure can... be a misleading indicator of income inequality itself".

These quantitative methods of measuring horizontal equity represent only a sample of those available to academics seeking to evaluate the impact of alternative tax proposals. Because the focus of this monograph is not dependent on any specific tax proposal(s), we do not use these techniques in our examination and discussions of the tax policy issues that follow.

SUMMARY

The publication of academic studies of horizontal and vertical equity will persist. However, we agree with Hyman (1987) – these studies possess considerable potential for subjectivity.

Studies of horizontal equity, vertical equity, explicit taxes, and implicit taxes remain useful as frameworks for academic debate. They represent a starting point or forms of quantitative nomenclature and are vehicles through which the equity of anticipated (and imposed) tax law changes can be debated.

PART 2

FAMILY TAXATION

CHAPTER 6
EXEMPTIONS AND STANDARD DEDUCTIONS

... the personal exemptions perform at least three functions:

(1) They exclude from the tax base entirely the earnings of the lowest strata of income recipients.
(2) They provide a factor of graduation, especially important for the income class associated with the first bracket in the scale; here all the graduation is supplied by the personal exemption.
(3) They differentiate among families according to the number of dependents and provide a special concession to the aged and the blind (Groves, 1963, 18).

INTRODUCTION

Groves (1963) provided a superior analysis of the economic and policy issues relating to personal and dependents exemptions and the standard deduction. His study also remains useful and relevant as a foundation for contemporary researchers of the related progressivity and equity issues addressed in Chapters 4 and 5, respectively.

In this chapter, our discussion focuses primarily on the evolution of personal and dependency exemptions as they relate to the U.S. individual federal income tax. Standard deductions represent a related topic and are also addressed. Our framework extends the work of Brozovsky and Cataldo (1994, 168–169). We also include an examination of changes in the Canadian counterpart, where relevant.

U.S. personal and dependency exemption amounts are inflation-indexed. We do not anticipate a change in this area of U.S. individual federal income tax policy. Instead, we expect to see a continuation and expansion of the use of targeted tax credits, in particular, the continued growth of the earned income tax credit, a topic covered in Chapter 7.

49

THE PERSONAL EXEMPTION (AND THE STANDARD DEDUCTION)

The original U.S. Federal tax law (1913) provided a $3,000 personal exemption for a single person and a $4,000 personal exemption for a married couple. There was no separate deduction or dependency exemption until the 1917 tax year, and there was no standard deduction until the 1944 tax year (see Chapter 2).

Appendix Table 6-1 contains a detailed summary of the entire history of U.S. personal exemptions, dependent allowances, and standard deduction amounts for single (SGL) and married taxpayers filing jointly (MFJ).[86] These dollar amounts were periodically increased or decreased by statute.

Beginning with the 1944 tax year, the standard deduction was introduced at a *variable* (Brozovsky & Cataldo, 1994, 168–169) rate of 10% of the taxpayer's AGI, to a maximum of $1,000. Beginning with the 1964 tax year, the standard deduction took a *semi-variable* (Brozovsky & Cataldo, 1994, 168–169) form. A standard deduction floor and ceiling were established at $200 and $1,000 respectively, and within this range the standard deduction was available for 10% of the taxpayer's AGI (1964–1970). These ceilings, floors, and percentages were increased to a maximum of 16% of AGI through the 1976 tax year (see Appendix Table 6-1).

Beginning with the 1977 tax year, the standard deduction was *fixed* (Brozovsky & Cataldo, 1994, 168–169). Inflation-indexing was applied for the 1985–1987 tax years and reapplied to a revised or increased standard deduction base for the 1988 and subsequent tax years. Beginning with the 1991 tax year, standard deduction phase-outs were applied to high-income taxpayers.

Appendix Table 6-2 contains a summary of the Canadian counterpart for personal exemptions (1917–) and dependent allowance deductions (1918–1941 and 1947–1992), the latter having been replaced with a tax credit from 1993 onwards.

One view of the personal exemption amount is that it should approximate some minimal standard of living maintenance level (Groves, 1963, 26). However, the amount of $2,750 available for the 1999 tax year for a single taxpayer, is below that available for the 1913 tax year (at $3,000). This suggests that this notion has been discarded. . . perhaps in favor of the earned income tax credit (EITC) (discussed in Chapter 7).

Groves examined five different "exemption" techniques: (1) the *continuing* exemption, (2) the *initial* exemption, (3) the *vanishing* exemption, (4) the *tax credit*, and (5) the *percentage-of-income* allowance. In the U.S., these techniques (or hybrids) have been used for both personal/dependent exemptions and the standard deduction for contemporary non-itemizer taxpayers. These techniques are discussed below.

(1) The Continuing Exemption

The continuing exemption, in its pure form, is the one most commonly used in the U.S. (1913–1990). It is adjusted through the introduction of personal exemption phase-outs for high-income taxpayers (see Appendix Table 2-1; 1991–).

The continuing exemption/deduction provides a tax benefit which is of greater value to higher-income taxpayers, who are subject to higher progressive tax rates/brackets. For example, two taxpayers receiving the same personal exemption/deduction of $1,000 may receive a tax benefit of $150, if in the 15% FIT bracket, or a tax benefit of $280, if in the 28% FIT bracket. Since the higher tax rate/bracket taxpayer has greater ability-to-pay, the higher tax benefit would appear to be inequitable.

If the purpose of the personal and dependent exemptions/deductions is to provide for some minimal level of subsistence, as recent legislation for the phase-out of these personal exemptions for high-income taxpayers would suggest (see Appendix Table 2-1; 1991–), it should either be: (1) eliminated for taxpayers above some operationally agreeable definition of the poverty level, and/or (2) the phase-out of personal exemptions should be introduced into the tax rates structure at a much lower AGI level.

For the 1999 tax year, the phase-out of personal exemptions began at the $126,600 (SGL), $189,950 (MFJ), $94,975 (MFS), and $158,300 (HH) AGI levels. If one is willing to presume that taxable incomes approximate 87.5% of AGI (IRS SOI, 1996, 3), taxpayers are not adversely affected by the personal exemption phase-out until they reach the 31% (for the HH and SGL filing status) or 36% FIT bracket (for the MFS and MFJ filing status).

(2) The Initial Exemption

The initial exemption represents a feature comparable to contemporary U.S. individual income tax policy. Here, the objective is to confine all or most of the tax benefits/savings to taxpayers in the lower income classes/brackets. Again, the introduction of the personal exemption phase-out (see Appendix Table 2-1; 1991–) represents an evolutionary move in this direction. However, as previously discussed, the contemporary phase-out of personal exemptions does not achieve the differentiation or additional progressivity for families in the bottom FIT brackets.

(3) The Vanishing Exemption

The vanishing exemption is the form presently used in the U.S. It is facilitated through the use of phase-outs (see Appendix Table 2-1; 1991–).

The "vanishing" feature can also take the form of a tax credit. Canada introduced a vanishing, refundable child tax credit, beginning with the 1978 tax year, for children eligible for family allowance (CTF, 1986–1987, 7:7).

(4) The Income Tax Credit

Beginning with the 1998 tax year, the U.S. provided for a child tax credit of $400 ($500 for 1999–) for dependent children under age 17 at the close of the calendar year. This child tax credit possesses "vanishing" features very similar to the personal exemption phase-outs (see Appendix Table 2-1), as the allowable credit is reduced by $50 for each $1,000 (or fraction) of modified AGI above $110,000 for joint filers, $75,000 for unmarried individuals, and $55,000 for married taxpayers filing separate returns (for the 1998 tax year only). The income tax credit is also used by some U.S. states.

Canada converted to use of a non-vanishing tax credit based on their perception of *equity* considerations (1988–):

> Under the 1987 system, an individual's "income" was reduced by personal exemptions and other deductions to which he was entitled in determining his "taxable income". Under the reformed system, most of the exemptions and deductions have been converted to non-refundable tax credits, at the rate of 17% of basic amounts.
>
> The credit system is based on the assumption that a certain amount of income should be effectively exempt from tax at the first rate of 17%. All taxpayers receive the same tax savings, in absolute terms. The credits are indexed, beginning in 1989, at the annual rate of change in the consumer price index (CPI) less three percentage points (CTF, 1990, 7: 4 and 7: 6).

The Canadian use of a tax credit instead of a tax deduction is equivalent to a vanishing deduction that never completely vanishes, avoids the notch problem associated with the initial exemptions approach, and reduces the progressive nature of the tax system to which it applies.

(5) The Percent-of-Income Exemption (Standard or Itemized Deduction)

The percent of income (AGI) exemption, in the form of a variable standard deduction, was used in the U.S. individual Federal income tax system from 1944–1976 (see Appendix Table 6-1). For the 1944–1963 tax years, it took the form of a 10% of AGI standard deduction to a *maximum* (or ceiling) amount of $1,000. For the 1964–1970 tax years, a *minimum* standard deduction feature (or floor), of $200, was added. From 1971–1976 the percentage was increased from 13% to 16%, with both minimum and maximum deduction amounts, later providing separate measures for single and married taxpayers (1975 and 1976).

Presently, a percent-of-income (AGI) approach is applied to high-income U.S. taxpayers with itemized deductions. The "3%/80% rule" involves the reduction of itemized deductions when a taxpayer's AGI exceeds an inflation-adjusted "applicable amount" (1991–). For the 1999 tax year, otherwise allowable itemized deductions are reduced by the lesser of:

- 3% of the excess of AGI above $126,600 ($63,300 if MFS),[87] or
- 80% of the itemized deductions otherwise allowable for the tax year.

Figure 6-1 summarizes the five different exemption techniques addressed by Groves in 1963. Also indicated are the applicable tax years to which they applied in the U.S. individual federal income tax system.

Contemporary trends are for the phase-out of the dependency exemption (1991–), a variation in form, but not in substance, from Groves' *initial* and *vanishing* exemptions, and also for a phase-out of the child tax credit (1998–) or Groves' *income tax credit*. Unlike the Canadian system – with no phase-outs – the U.S. system of phase-outs is designed to prevent high-income taxpayers from benefiting from these exemptions and credits, which have been designed to assist lower income taxpayers.

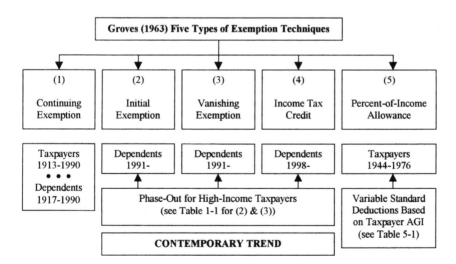

Fig. 6-1. Exemption Techniques Used in the U.S.

THE REASONING BEHIND EXEMPTIONS

McIntyre and Steuerle subscribe to the *traditional view* that an individual Federal income tax system should fairly distribute tax burdens, regardless of family circumstances. They examined family taxation and the *ability to pay* issue under three fundamental contemporary tax policy reform proposals (1996, 13):

(1) The Armey/Shelby flat tax (Flat Tax): Individuals and businesses would be taxed at a flat rate of 20% for the first two years; and 17% thereafter. This proposal would have replaced the individual and corporate income and gift and estate taxes. This plan would have resulted in a loss of Federal tax revenues ($30 to $138 billion);

(2) The Nunn/Domenici tax system (USA Tax): Individual Federal income taxes would be converted to a progressive consumption tax; the corporate Federal income tax would be replaced with a flat value-added tax (VAT). The FICA payroll tax would be (implicitly) replaced, in the form of a tax credit. Individual Federal income tax rates of 19%, 27%, and 40% would be established and reduced (in stages) to 8%, 19%, and 40% (after five years). This plan was found to be *revenue neutral* by the Joint Committee on Taxation; and

(3) The Gephardt 10-Percent Tax (10% Tax): An individual Federal income tax rate of 10% would apply to about 75% of taxpayers. High-income taxpayers would be subjected to progressive rates of 20%, 26%, 32%, and 34%. Deductions would be reduced. Treasury Department estimates found this plan to be *revenue neutral*.

McIntyre and Steuerle cite several examples of the current Federal income tax system that take into consideration family economic circumstances (1996, 11–12):

(1) The dependency exemption (discussed below);
(2) The earned income tax credit (discussed in Chapter 7);
(3) The child and dependent care credit (addressed in Chapter 6);
(4) The marital joint filing and income splitting rules (not discussed);
(5) The deduction for alimony and non-deductibility of child support (not discussed);
(6) The differing household-based standard deduction amounts (discussed in Chapter 6); and
(7) The special Federal income tax rate schedule for heads of households (addressed in Chapter 6).

In addition, both republicans and democrats proposed the addition of

(8) The child dependency credit.

The dependency exemption represents a tax burden adjustment based on family size and therefore *ability to pay*. It is based on the theory that the income used for the basic needs of children should be exempt from tax.

The dependency exemption is: (1) a major mechanism for adjusting tax burdens for the support of children, (2) the only mechanism providing tax benefits to all middle-income families with dependent children, and (3) phased-out for high-income taxpayers (1987–), raising the effective tax rate above the 39.6% tax rate schedule ceiling for some high-income taxpayers (consistent with *welfare* considerations, but inconsistent with the *tax policy* consideration).

Pechman examined the *personal exemption* and *standard deduction* amounts (1939–1989), their contribution to progressivity and as an administrative device to remove low-income taxpayers from the tax rolls (1987, 81). He cited the Census Bureau's annually published official poverty-line estimates, based on the income level required to maintain an adequate diet, as the standard criteria used by the U.S. federal government (1987, 83). His comparison of "financial needs" to combined income tax exemption plus standard deduction ratios for the 1989 tax year suggested that the per capita exemption was: (1) too liberal for dependents, and (2) too small for single persons.[88]

He discussed the merits of using income tax credits to supplement or replace personal exemptions and standard deduction amounts, but cited the historical use of this (supplemental) system (used to adjust personal exemptions) over the 1975–1978 period ($30 in 1975 and $35 for 1976–1978). The practice was eliminated due to its complexity and ineffectiveness (Pechman, 1987, 85). The merits of inflation-indexing (Pechman, 1987, 115) and the "vanishing exemption" (Pechman, 1987, 85), also known as the contemporary "phase-out" of personal exemptions for high-income taxpayers, were also discussed and represents our present U.S. focus on the *ability to pay* concept.

Personal and dependent exemptions are intended to provide a mechanism for the elimination of low-income taxpayers, with little or no taxpaying capacity, from the administrative and/or economic requirements of paying income taxes. The appropriate amount of the exemption constantly changes, as price levels do, and is a topic of frequent inquiry. Should the personal exemption represent the amount necessary for basic needs? And how are "basic needs" to be operationally defined? Do they include food, shelter, clothing, medical (and dental) insurance, etc.? And even assuming that some agreement might be reached on the amount of the personal exemption or deduction for a single taxpayer, should this amount be doubled in the case of married taxpayers with no dependents, tripled in the case of a household with one dependent, quadrupled in the case of a household with two dependents, etc.? Or should economies of scale (e.g. two can live almost as cheaply as one) be applied?

Presently, the U.S. FIT system, where the household is the unit of taxation, provides for equivalent individual and dependent exemption deductions (e.g. $2,750 each for the 1999 tax year; see Appendix Table 6-1). For high-income taxpayers, these exemptions are phased out (see Appendix Table 2-1). However, the phase-out thresholds are quite high, affecting taxpayers with taxable incomes (or AGIs) far above poverty levels (e.g. phase-out began at a 1999 tax year AGI of $126,600 for single taxpayers). Similarly, the phase-out of itemized deductions affects only relatively high-income taxpayers (e.g. phase-out began at a 1999 tax year AGI of $126,600, $63,300 if MFS).

U.S.-CANADIAN COMPARISONS OF PERSONAL AND DEPENDENT EXEMPTIONS

Since 1987, Canada has replaced the personal exemption with a tax credit at the lowest FIT rate/bracket of 17%. Canadians were motivated by a desire to achieve equity. Canadian policy and practice suggests that equity is achieved under a condition where the taxpayer's tax bracket does not affect dependent-based tax savings.

Figure 6-2 provides a graphical comparison of U.S./Canadian personal exemptions for single taxpayers without dependents. Figure 6-3 provides a similar comparison for married taxpayers without dependents. In both cases, contemporary Canadian personal and spousal exemption tax credits have been restated in the form of deductions, to facilitate comparison.

The U.S. individual FIT system began in 1913. The Canadian individual FIT followed four years later, and first became effective for the 1917 tax year.

In the case of the single taxpayer without dependents (see Fig. 6-1), U.S. and Canadian personal exemptions, with only two exceptions and ignoring currency differences, were exactly the same from tax year to tax year, from 1917–1941. Beginning with the 1942 tax year, the Canadian personal exemption amount exceeded that of the single U.S. taxpayer counterpart. Canada was the first to apply inflation-indexing to the personal exemption, beginning with the 1974 tax year. The U.S. adopted this practice for the 1985, 1986, and 1988 and subsequent tax years. However, Canada has not increased the personal exemption deduction since the 1992 tax year, though inflation-indexing still applies for amounts in excess of 3%.

In the case of the married taxpayer without dependents (see Fig. 6-2), the U.S. provided for larger personal exemptions through the pre-/early-World War

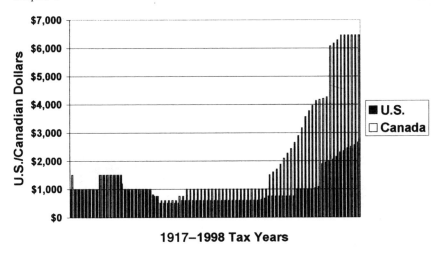

Fig. 6-2. A Comparison of U.S./Canadian Personal Exemptions. Single Taxpayers – No Dependents.

N = 82
r = 0.860 (Pearson correlation coefficient)
$R^2 = 73.9\%$
Overall F-statistic = 226.78; p-value = 0.000

Note: U.S. & Canadian dollars have not been inflation-adjusted or adjusted for differentials in purchasing power between the two currencies.

II period (1940). These measures remained constant for three years (1941–1943), and then a pattern similar to that for single taxpayers emerged. The Canadian personal exemptions outpaced those available in the U.S.

   In viewing these differences, it is important to recall that the unit of taxation in the U.S. in the *household*. The contemporary U.S. system of FIT rates provides different tax rate schedules for single (SGL), head of household (HH), and married (MFJ) taxpayers, plus a standard deduction. In Canada, the unit of taxation is the *individual*, a standard deduction is not available, and only one tax rate schedule applies to all taxpayers.

   Deductions for dependents were generally lower for Canadians (1923–1931) or equivalent to those amounts available in the U.S. (1932–1941) through the WWII period. Since 1942, the U.S. has provided for increasingly larger deductions for dependents, though these deductions do not take the Canadian child credit into consideration. The Canadian system, unlike that of the U.S., has always provided for differences based on age or familial relationships.

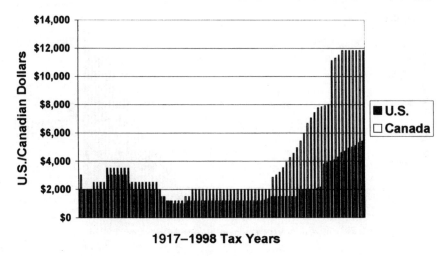

Fig. 6-3. A Comparison of U.S./Canadian Personal Exemptions. Married Taxpayers –
No Dependents.

N = 82
r = 0.783 (Pearson correlation coefficient)
$R^2 = 61.3\%$
Overall F-statistic = 126.92; p-value = 0.000

Note: U.S. & Canadian dollars have not been inflation-adjusted or adjusted for
differentials in purchasing power between the two currencies.

Figure 6-4 provides a comparison of U.S. and Canadian dependent exemptions
(1917–1998).

   Figures 6-5 and 6-6 provide stacked area graphs of U.S. (1913–1998) and
Canadian (1917–1998) personal and dependent ratios, where the personal
exemption available to the single taxpayer without dependents is set to 1.0. In
each case, the personal and dependent exemptions were developed as a ratio of
single taxpayers without dependents (SGL), married taxpayer filing jointly
(U.S. only) and without dependents (MFJ), and MFJ with one dependent child
(MFJ + 1) or SGL:MFJ:MFJ + 1.

   As Fig. 6-5 illustrates, the post-WWII era (1944–1998) has consistently
employed a 1 : 1 : 1 ratio of SGL:MFJ:MFJ + 1 personal and dependent
exemptions in the U.S. As Fig. 6-6 illustrates, the Canadian system provided
for equivalent personal exemption amounts for married taxpayers through the
1971 tax year, but has provided for lesser amounts for the "MFJ" counterpart.
Again these exhibits do not reflect the Canadian child tax credits.

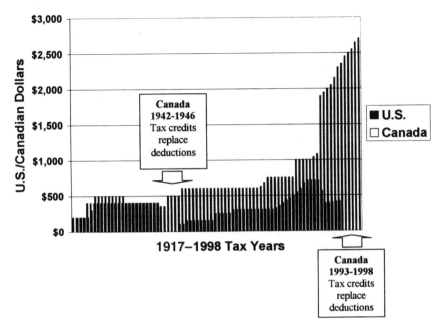

Fig. 6-4. A Comparison of U.S./Canadian Dependent Exemptions.

N = 82

Note: U.S. & Canadian dollars have not been inflation-adjusted or adjusted for differentials in purchasing power between the two currencies.

SUMMARY

The U.S. system of phasing out the personal and dependency exemptions and itemized deductions (1991–) and the phasing-out of the recently enacted child tax credit (1998–) reflects the contemporary (and perhaps more liberal) trends toward the retention of a progressive tax system of taxation (Chapter 4) with personal exemption and standard deductions. For example, Vice-President Al Gore, during the 2000 Presidential campaign, proposed an increase in the standard deduction for married couples as a mechanism to provide marriage tax penalty (discussed in Chapter 8) relief.

In the U.S., the phase-outs prevent high-income taxpayers from disproportionately benefiting from these deductions and credits. Canada, on the other hand, views the provision of dependency tax credits, at a fixed tax bracket equivalent of 17%, as a more equitable approach.

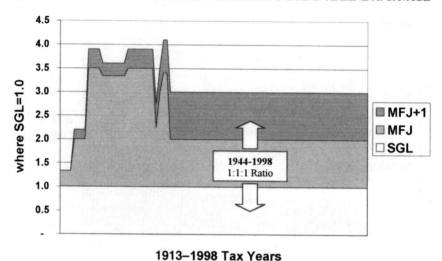

Fig. 6-5. U.S. SGL:MFJ:MFJ + 1 Dependent Exemption Ratios.

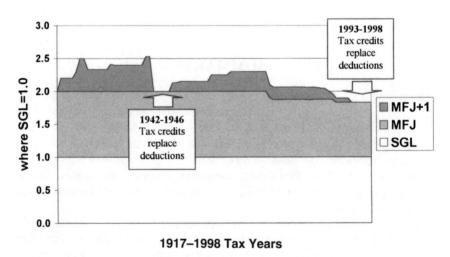

Fig. 6-6. Canadian SGL:MFJ:MFJ + 1 Dependent Exemption Ratios.

The U.S. child tax credit (1998–) reflects a trend toward the use of supplemental tax credits in lieu of additional or increased tax deductions. However, during the 2000 Presidential campaign, George W. Bush proposed that an additional personal exemption be made available for caregivers of long-term care. He also proposed that the child tax credit be doubled from $500 to $1,000. Vice-President Al Gore also favored expansion of the child and dependent care tax credits.

The initial ceiling of $400 for the child tax credit (1998–) is the same as that amount initially available (1975) for the earned income (tax) credit (EI(T)C). The EIC is subject to phase-in, flat, and phase-out ranges. It is the topic of the next chapter.

CHAPTER 7
THE EARNED INCOME TAX CREDIT

The *Liberal* envisions the EIC as a supplement to, not a replacement for, welfare. *Conservatives*, on the other hand, envision the EIC as a replacement for welfare for non-elderly, able-bodied people (Hoffman & Seidman, 1990, 3).

INTRODUCTION

The contemporary earned income (tax) credit (EI(T)C) began with the 1975 tax year. It is a form of "workfare", requiring that taxpayers generate *earned income* in order to benefit from this refundable credit.

The contemporary EIC provided for an initial maximum refundable tax credit of $400 for the 1975 tax year. This maximum has increased to $3,816 (for a taxpayer with two or more dependents) for the 1999 tax year. This growth is likely to continue.

The contemporary EIC is the third in a series of tax-related measures. All three were examined by Cataldo (1995) and are briefly summarized below.

CONTEMPORARY AND PREDECESSOR EARNED INCOME CREDITS

The contemporary EIC is summarized in Appendix Table 7-1. It commenced in the 1975 tax year and applied to more than 7.5% of the individual federal income tax returns filed. For the 1996 tax year, the EIC applied to more than 16% of returns filed.

The first EIC was applied as a credit against tax (1923–1931). The second, unlike the first and contemporary tax *credit*, was applied against net (taxable) income (1934–1943). Therefore, it was very similar to a tax *deduction* and, in fact, evolved into the early version of the contemporary U.S. standard deduction (see Chapter 6). The third (and contemporary) EIC takes the form of a refundable tax credit (1975–). This framework was presented and examined in Cataldo (1995). Descriptive statistics for all three EICs, using IRS SOI-based measures, are provided in Appendix Table 7-2.

Cataldo further segregated the contemporary EIC into three phases: Phase I (1975–1990) provided for the implementation of the EIC, where the maximum EIC doubled, rising from $400 to $953; Phase II (1991–1993) provided for supplemental health care and newborn components of the EIC, but was later abandoned; and Phase III (1994–) provided for the inclusion of low-income, no dependent taxpayers. The EIC credit rates, "flat" ranges, and maximum dollar amounts for the contemporary EIC are summarized in Appendix Table 7-2.

EIC-BASED POLICY ISSUES

> ... the EIC now often seems to be the policy of choice for a wide range of problems ranging from the minimum wage ("increase the EIC instead") to child care ("adjust the EIC for family size") (Hoffman & Seidman, 1990, 1).

The initial intent of the 1975 EIC was to offset the *regressive* nature of Social Security payroll taxes for low-income workers with children. This foundation has been retained in the case of the low-income, no dependent taxpayer, as is suggested by the 7.65% EIC rate (a rate which is equivalent to the employer/employee FICA tax contributions) for this taxpayer group (see Appendix Table 7-2 for 1994–).

The EIC versus Aid to Families with Dependent Children (AFDC)

Unlike AFDC, EIC benefits are provided only to families with positive *earned income*. Also unlike AFDC, the EIC provides benefits to both *married* and single parent families. Therefore, unlike AFDC, the EIC supports both *work* and *marriage* (Hoffman & Seidman, 1990, 2).

The EIC versus the Minimum Wage

Unlike increases in the minimum wage, the EIC is well-targeted and based on a household's total income, rather than an individual's hourly wage. Increases in the minimum wage are poorly targeted, benefiting teenagers in middle- and high-income families. Increases in the minimum wage will generate involuntary unemployment in areas where the legal minimum wage exceeds the market wage. Therefore, the EIC is more *antipoverty effective* and does not generate *involuntary employment* on the labor supply (Hoffman & Seidman, 1990, 2).

The EIC and Child Care

Beginning with the 1991 tax year, the EIC was modified to incorporate family size into the computation of benefits. For example, for the 1991 (1999) tax year a family with one dependent child could generate a maximum EIC of $1,192 ($2,312). For the same tax year(s), a family with two dependent children could generate a maximum EIC of $1,235 ($3,816) (see Appendix Table 7-2).

MECHANICS OF THE EIC

The EIC is intended to provide an economic incentive for the taxpayer to generate earned income (EI) or to "work". Figure 7-1 illustrates the mechanics of the contemporary EIC-based phase-in, flat, and phase-out ranges for single (no dependent), one dependent, and two (or more) dependent households for the 1999 tax year. This figure (and the discussion that follows) was generated using the information contained in Appendix Table 7-1.

Phase-in Range

The phase-in range of the EIC illustrates the positive "workfare" features of the EIC. As EI increases, the low-income taxpayer is rewarded with a higher percentage of EIC (see Fig. 7-1). For example, the qualifying single taxpayer

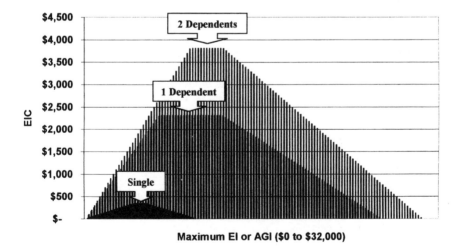

Fig. 7-1. The 1999 Earned Income Credit: Phase-In, Flat & Phase-Out Ranges.

(0 dependents in Appendix Table 7-1 for the 1999 tax year) receives an EIC of 7.65% to a maximum EI level of $4,530. This results in a refundable EIC of $347. If this same taxpayer had EI of only $1,000, the maximum refundable EIC would be only $76.50 ($1,000 *multiplied by* 7.65%). Therefore, throughout the phase-in range, this taxpayer is provided with an economic incentive to work and generate wage or salary income.

Flat Range

The flat range is that EI range over which the refundable EIC *remains* maximized. As the taxpayer's EI increases throughout the flat range, EIC benefits remain maximized. For example, the qualifying taxpayer with one dependent (see Appendix Table 7-1 for the 1999 tax year) receives an EIC of 34% to a maximum EI level ranging from $6,800–$12,460. This results in a refundable EIC of $2,312. Therefore, throughout the flat range, this taxpayer retains the economic incentive to work and generate wage or salary income.

Phase-out Range

The phase-out range of the EIC illustrates a potential weakness in the contemporary EIC. As EI increases beyond the flat range, the taxpayer's refundable EIC benefits are reduced. For example, the qualifying taxpayer with two (or more) dependent children (see Appendix Table 7-1 for the 1999 tax year) receives a maximum EIC of 40%, or a maximum of $3,816, which is phased-out at a rate of 21.06%. Complete phase-out is achieved when the taxpayers EI reaches $30,580. Therefore, after this taxpayer has generated wage or salary income of $12,460, he/she would lose refundable EIC benefits at a rate of $21.06 for each additional $100 in wage or salary income.

CHOOSING LABOR OR LEISURE

The opportunity cost of *leisure* is the salary or wage *income* given up to enjoy leisure. As income rises the opportunity cost of leisure also rises. Therefore, to the extent that the EIC raises incomes for those engaged in labor, an increasing percentage of taxpayers may choose labor over leisure. Economists examine the trade off between labor and leisure in terms of *income* and *substitution* effects and the *utilities* generated by each.

Income Effects

Generally, the income effect of the EIC has no effect on taxpayers already working. For these taxpayers, the EITC will increase income for the phase-in, flat, and phase-out ranges. Generally, the EITC tends to encourage labor and discourage leisure throughout the *phase-in* range. The income effect of the *flat* range does not promote or discourage labor. The income effect of the *phase-out* range of the EIC provides an economic *disincentive* for potential EIC recipients to choose labor over leisure.

Substitution Effects

Generally, the substitution effect of the EIC encourages work throughout the *phase-in* range. The substitution effect of the flat ranges does not promote or discourage labor. The substitution effect of the *phase-out* range of the EIC provides an economic *disincentive* for potential EIC recipients to choose labor over leisure.

Utility

Generally, income and substitution effects are a function of *utility*. For this reason, the labor supply curve is frequently depicted as backward-bending. Therefore, after some level of income is achieved, the labor supply will decline. This is because the affluent worker will choose more hours of leisure as his/her earnings increase.

Affluence is a relative term and a function of *individual utility*. Therefore, for those taxpayers with a relatively strong desire (or utility) for leisure, the *phase-in* range of the EIC may continue to provide adequate "workfare" incentives. However, this same group of taxpayers would be greatly influenced by the *phase-out* range of the EIC, which will produce an even stronger disincentive for the selection of labor over leisure.

ADMINISTRATIVE DIFFICULTIES

The EIC is viewed favorably by both liberals and conservatives. However, the program remains ineffective in its current form for the following reasons (Cataldo, 1995; Scholz, 1994; Yin & Forman, 1993; Steuerle, 1993; Kirchheimer, 1993; and Holt, 1994):

(1) Some EIC-qualifying taxpayers are not otherwise required to file an income tax return and may not receive the benefits they are entitled to;

(2) The advanced payment option, the most effective means of EIC benefits delivery, is almost completely ineffective;
(3) Self-employed taxpayers may fraudulently report work in the phase-in range to maximize EIC benefits;
(4) Recipients of the advanced EIC payment option may, at year-end, prove to be ineligible;
(5) Income variability, multiple employers, and two-earner households electing the advanced EIC option, may experience EIC overpayments that must be repaid; and
(6) The contemporary EIC, like its welfare predecessors, provides economic disincentives for marriage.

SUMMARY

The contemporary earned income (tax) credit has been in effect for over a quarter of a century. Viewed as both a supplement and a potential replacement for welfare, the EIC remains popular and continues to be expanded. Both political parties favor use of the EIC, therefore, we expect this expansion to continue.

For academic researchers and legislators alike, the future of the EIC will be dependent on the ability to achieve greater success in solving existing administrative problems. Perhaps the most important weakness is that associated with the advanced payment option. Taxpayers qualifying for the EIC do not benefit by receiving *annual, lump sum* payments. Refinement of the advanced payment option would spread these benefits over the entire tax year, providing periodic *paycheck supplements* to these low or moderate income taxpayers.

CHAPTER 8
THE MARRIAGE TAX

Delaying a November or December wedding until next year could mean especially large tax savings for many two-income singles whose incomes are roughly the same . . . Many other people, such as unmarried couples with only one income, might be able to cut their . . . tax bill by getting married this year (*Wall Street Journal*, November 17, 1999, 1).
. . . candidates from both political parties vie to position themselves as champions of the American family (McIntyre & Steuerle, 1996, 11).

INTRODUCTION

The U.S. income tax system has never been marriage neutral, despite literature claims that our system of individual federal income taxation was marriage neutral prior to income splitting in 1948 (Alm & Whittington, 1993) or prior to the post-1970 development of separate rate schedules for married taxpayers (Neff, 1990; Rosen, 1987). However, marriage tax non-neutralities have been characterized as "small" (Rosen, 1988) prior to the post-1970 period.

This chapter is organized into two broad sections. The first section examines the entire history of the marriage tax. The focus of this section is on what has been referred to as the *rate effect* and the *base effect* components of the marriage tax. As this monograph goes to press, legislation that would eliminate rate and base effects (Cataldo, 1996) is under consideration. The second section examines recent historical marriage and divorce rates in the form of a marriage-to-divorce (M-T-D) ratio. The focus of this exploratory examination is on the post-1970 period (1991–1997), providing a comparison between the U.S. and Canada. Recall that the Canadian individual federal income tax system taxes the *individual*. Therefore, unlike the American system, the Canadian individual federal income tax system is generally marriage tax neutral.

THE MARRIAGE TAX PENALTY AND MARRIAGE TAX BONUS: *RATE* AND *BASE* EFFECTS

The marriage tax penalty/bonus (MTP/MTB) refers to that situation where a legally married couple using

(1) Married, filing joint (MFJ) or
(2) Married, filing separate (MFS)

tax rate schedules pays higher/lower federal income taxes than what would be paid by two single taxpayers with the same level of income. The MTP has historically been viewed as originating from a "base" effect and the more significant post-1970 "rate" effect (Fox, 1988).

The *base* effect (IRC Section 63) MTP is a function of the differing standard deduction amounts statutorily available to single and married taxpayers (see Appendix 1A). For example, for the 2000 tax year, non-itemizing single taxpayers will receive a standard deduction of $4,400 each, while married taxpayers filing jointly will get a 2000 standard deduction of $7,350 per return. Therefore, the 2000 MTP associated with the base effect is equal to $1,450 (i.e. [$4,400 × 2] – $7,350) multiplied by the taxpayer's marginal tax rate (e.g. $218 for taxpayers in the 15% tax bracket, where the $1,450 difference is *multiplied by* 15%). The base effect generates a MTP for relatively low-income, non-itemizer taxpayers.

The *rate* effect (IRC Section 1) MTP is a function of the differing tax rate tables available to single and married taxpayers for post-1970 tax years. For example, for the 2000 tax year 15% and 28% tax brackets, breakpoints for single taxpayers occur at a taxable income level of $26,250 each. The related 2000 breakpoint for married taxpayers, filing jointly, occurs at a taxable income level of $43,850 per couple. Therefore, the 2000 maximum MTP associated with 15% and 28% brackets provides for a rate effect of $2,422 (i.e. [[$26,250 × 2] – $43,850] × 28%).

Other MTPs have been addressed in professional journals (Tilt & Spencer, 1983; Jagolinzer & Strefeler, 1986; Brozovsky & Cataldo, 1993). Generally, these result from AGI-based limitations and the tax treatment differentials are based solely on the taxpayer's filing status.[89]

Historical Evidence – 1914–1996

In the United States, during the worst depression year (1932) the marriage rate was cut to one-fifth below the prevailing level of the 1920s, and in the immediate postwar period

(1946–47) the rate was up more than that amount above the level of the early 1940s (Carter & Glick, 1970, 386).

Between (1890) . . . and 1960, the divorce rate for married persons underwent a three-fold increase, with most of the rise taking place by 1920. It rose very rapidly, but temporarily, during and soon after World War II. In the postwar period, the divorce rate moved downward until 1955 then rose by 1965 to about the same level as in 1949 (Carter & Glick, 1970, 387).

Alm and Whittington (1993) noted that a "(d)eparture from marriage neutrality developed with the adoption in 1948 of income splitting for married couples". This misconception has, from time to time, surfaced in the literature (e.g. Neff, 1990; Mitchell, 1989). Our current system of individual federal income taxation has never been "marriage neutral" (Brozovsky & Cataldo, 1994). Rosen (1988), in his response to McIntyre's (1988) critique of Rosen's 1987 study, characterized historical non-neutralities as "small". However, when these small non-neutralities were adjusted for inflation their significance became apparent. And when shown in light of their impact on the average taxpayer's total tax bill, early (pre-WWII) non-neutralities were of greater significance than more recent (post-1970) non-neutralities.

We define marriage tax non-neutralities, based on income allocations of selected levels of AGI, as differences between single taxpayers versus married taxpayers under a no dependent, non-itemizer scenario. This approach differs from that of Alm and Whittington's (1993) or Rosen's (1987) methods of marrying average single taxpayers or divorcing all married taxpayers, respectively. The *marriage-based* approach employed by Alm and Whittington is not appropriate for the early history of the income tax, not only because of the lack of adequate and consistent detailed data, but also because a large (though declining) fraction of female citizens did not file tax returns prior to WWII. Alternatively, Rosen's (1987) *divorce-based* approach cannot be used due to the lack of comparable historical data. A different approach is necessary to allow for the application of a consistent methodology throughout the history of our current system of individual federal income taxation.

Drawing from the terminology employed by Fox (1988), we focus only on the rate and base effects in the generation of MTPs and MTBs. Rosen's introduction of a third element, the EIC (see Chapter 7), was severely criticized by McIntyre (1988). Therefore, the calculations contained in this chapter exclude the EIC element as well as the possibility of any related head of household status-based effects. By using a no dependent scenario, the incorporation of head of household preferential rate (first established for the 1952 tax year) and base (first established for the 1988 tax year) effects are avoided. Rosen's (1987) divorce-based simulation incorporated head of household EIC, rate, and base effects into his computation of the average overall marriage tax penalty. Rosen (1987, 567–568), and other researchers of this topic, have associated the introduction of marriage tax penalties with the 1971 tax year imposition of separately developed tax rate tables for married and single taxpayers. However, marriage tax penalties have always existed for certain groups of taxpayers.

For example, for the 1914–1916 tax years marriage tax penalties appeared for average taxpayers. However, during this period, the average citizen did not file an income tax return. The first penalty affecting average citizens (i.e. the middle 50%) appeared in 1964. This is earlier than the 1971 tax year imposition of the separate tax rate tables frequently associated with the introduction of marriage tax penalties.

Historical detailed evidence clearly illustrates a lengthy history of marriage tax non-neutralities (1914–1996). These data represent an extension of the work of Brozovsky and Cataldo (1994).

Appendix Tables 8-1 and 8-2 provide detailed first quartile, weighted average, and third quartile AGI measure-based MTPs and MTBs, respectively. Related Appendix Figs 8-1 and 8-2 graphically depict the *maximum* MTPs and MTBs, respectively, after CPI adjustment to 1998 dollars. In support of the weighted average measure of central tendency, Appendix Table 8-3 provides both nominal and 1998 CPI-adjusted weighted average AGIs. These measures are graphically illustrated in Appendix Fig. 8-3.

Several features of the MTP and MTB warrant elaboration. First, the MTP was modest from 1918–1963 (see Appendix Fig. 8-1). It peaked in 1981, immediately prior to the establishment of the two-earner married couple deduction (see Appendix 3F). Second, MTBs have generally increased (see Appendix Fig. 8-2) over the same time period. Third, weighted average AGIs, the measure upon which these MTPs and MTBs were computed, have increased at a very modest CPI-adjusted rate (see Appendix Fig. 8-3).

Historical and Contemporary Evidence

Although non-neutralities were small in nominal terms, they were substantial when adjusted for inflation. They were also material when viewed as a percentage of weighted average AGI. And when compared with the actual tax bill paid by an average married couple, early non-neutralities were far more significant than those occurring after 1947 or 1970, which have both been recognized as starting points for marriage tax non-neutrality.

Figure 8-1 illustrates the 1998 CPI-adjusted range of maximum MTPs and MTBs from 1914–1996. Notice the relative magnitudes. The potential for marriage tax *bonuses* is significantly larger than that for MTPs.

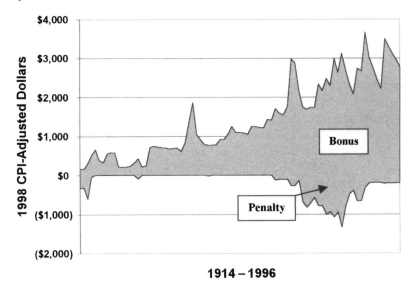

Fig. 8-1. Maximum U.S. Marriage Tax Bonus/(Penalty).

Limitations to Historical and Contemporary Analyses of MTPs/MTBs

The results presented in Fig. 8-1 reflect representations based on interquartile ranges. The calculations were based on discrete calculations of first quartile, weighted average, and third quartile-based AGI categories. However, our work with the calculations performed in this historical analysis suggests that significantly different graphical representations would not have resulted from the calculation of MTPs and MTBs for all possible AGI levels within this range and for all years represented by the period.

This analysis is confined to the development of marriage tax penalties/ bonuses relating only to base and rate effects under a non-itemizer, no dependent scenario. Although a number of other marriage tax penalty/bonus components exist today, Cataldo's (1996) findings, using 1989 SOI data, confirm that base and rate effects remain the primary components of the MTP. The post-1970 rate effect is clearly the greatest single contributor to the current marriage tax penalty.

Correcting Misconceptions in the Literature

Rosen (1987) referred to the pre-1948 U.S. personal income tax system as independent of marital status, while McIntyre (1988) noted that a return to this pre-1948 system would not result in marriage neutrality. Rosen (1988) concurred with McIntyre (1988) on the absence of pre-1948 "one hundred percent" marriage neutrality, but went on to state that "non-neutralities with respect to marriage were generally quite small", citing Pechman (1987). While the nominal non-neutralities may have been small, non-neutralities for average taxpayers were not insignificant. In fact, regardless of the method of adjustment, historical non-neutralities appear to have been quite significant.

Marriage tax bonuses persisted throughout the history of our current system of individual federal income taxation. Marriage tax penalties for the mid-range taxpayer, on the other hand, were infrequent in occurrence (1914–1917 and 1964–1970 tax years) and minor in both nominal and adjusted amounts prior to the (1971 tax year) application of substantively separate rate schedules for married and single taxpayers.

Non-neutrality with respect to marriage has never existed under the U.S. system of individual federal income taxation. Although marriage tax penalties were nearly non-existent through third quartile-based AGI levels for the 1917–1963 tax years, significant marriage tax bonuses were clearly present throughout the history of our current system of individual federal income taxation.

The next section examines U.S. and Canadian marriage-to-divorce ratios (M-T-Ds). This approach completely avoids the direct calculation and examination of the so-called marriage tax. Therefore, any methodological bias that might result from a: (1) divorce-based, (2) marriage-based, or (3) rate and base effect-based examination of the impact of the so-called *marriage tax* is also avoided.

THE MARRIAGE-TO-DIVORCE RATIO: A U.S./CANADIAN COMPARISON

Prior studies have found that tax laws penalizing/subsidizing married taxpayers can affect the *timing* of marital decisions in the United States, Canada, England and Wales. Only one of these studies (Alm & Whittington, 1995), an examination of the U.S. tax system (1947–1988), also concluded that MTPs may effect the *incidence* of marriage.

This section of the chapter provides a comparison of U.S. and Canadian marriage-to-divorce (M-T-D) ratios for the 1971–1997 tax years. M-T-D ratios (or *net marriages*) have declined, in both the U.S. and Canada, over this 27 year period. The Canadian tax system taxes the *individual* and does generally not have a MTP. The U.S. tax system taxes the *household* and/or provides for *joint* filing and generates both MTPs and MTBs.

Competing Studies and Their Conflicting Results

A single, U.S.-based study of net marriages (Alm & Whittington, 1995) concluded that the *incidence* of marriage and divorce were affected by MTPs. These results were inconsistent with a study by Sjoquist and Walker (1995) over a comparable period. However, both studies of the U.S. tax system, as well as those for Canada and England/Wales (Gelardi, 1996), found evidence to support the position that relevant tax law changes can affect the *timing* of marriage.

Unlike Alm and Whittington or Sjoquist and Walker, this present, exploratory study is restricted to marriage/divorce-based relations for the relatively stable, post-1970 period. This period excludes many of the post-WWII social changes, which were indirectly incorporated into models developed by researchers of U.S. MTPs. Furthermore, it relies only on the most frequently publicized and understood MTPs and does not require that taxpayers possess an awareness of their marginal tax rates.

Canadian M-T-D ratios have shown a far steeper decline than the U.S. counterpart (1971–1997). The Canadian trend toward a higher incidence of common-law marriage may partially explain these results and suggests that further study of taxpayer compliance with Canadian tax law (ITA 252 (4)) may be warranted. Canadian compliance-related findings may provide greater insights into the impact of MTPs on the *incidence* of marriage and ultimately supply an appropriate methodology for extensions to the United States and other countries.

The Marriage Tax

The marriage tax first evolved as an issue of concern in the United States with the introduction of separate tax rates tables for married and single taxpayers (1971–). It has been (anecdotally) suggested that the higher tax rates for married taxpayers, a form of MTP, can alter the decision to marry or remain married/divorce. Consequently, this MTP has been associated with a perceived decline (increase) in "family values" (and single parent households) in the U.S.

The MTP has been operationally defined as the incremental increase in tax paid by a married couple when compared to two single, cohabiting taxpayers at the same level of household income. Alternatively, the MTB reflects that situation where the same couple, if married, would pay a lower total tax. Therefore, the MTP/MTB can be illustrated, using Eq. (8-1), as follows:

$$[SGLTAX_1 + SGLTAX_2] - MRDTAX = MTB \text{ or } (MTP) \qquad (8\text{-}1)$$

where $SGLTAX_1$ and $SGLTAX_2$ are the taxes paid by single, but cohabiting taxpayers, and MRDTAX is the amount of tax that would be paid by the same taxpayers on the same amount of taxable income, if they were married.

Are Dependent-Related Effects Part of the MTP?

Some differences in the computations of MTP and MTB measures have resulted from the inclusion/exclusion of dependent-based tax deductions and credits. Generally, under U.S. law, comparisons between married and unmarried taxpayers with dependents yield higher income taxes, ceteris paribus, for the married taxpayers. This is partially due to the earned income credit (EIC; post-1974) and the more favorable standard deduction amounts and tax rates tables available to unmarried U.S. taxpayers qualifying for the "head of household" filing status.

This section isolates and examines only the "marital" decision component of MTPs and MTBs, as the origin of the phrase "marriage tax penalty" associated with the post-1970s imposition of separate tax rates tables would suggest. In U.S. terms, it addresses DINKs (double-income, no kids). This distinction is also made by McIntyre and Steuerle (1996, 25) in their U.S.-based examination of alternative, contemporary U.S. tax proposals.

Divorce-Based Descriptive Studies

Rosen (1987) used actual 1983 tax returns for married U.S. taxpayers to calculate the MTPs and MTBs resulting from hypothetical "divorces". McIntyre (1988) took exception to Rosen's failure to provide for a decomposition of his aggregate net measures into their separable components. His primary concern was the relative importance of the dependent-based U.S. EIC, a modified form of negative income tax, available to Rosen's simulated post-divorce taxpayers.

Feenberg and Rosen (1995) replicated Rosen's earlier work, but with actual tax returns from the 1989 tax year. This study again failed to provide for the decomposition of separable MTP and MTB components. Cataldo (1996) provided for this decomposition, as originally recommended by McIntyre (1988), using the same data base as Feenberg and Rosen. The results of Cataldo's analysis illustrated that U.S. MTPs primarily resulted from differing standard deductions ("base" effect) and the different post-1970 tax rates tables ("rate" effects) used by married and single taxpayers, both in terms of frequency and magnitude.

Marriage-/Divorce-Based Predictive Studies

Additional contemporary studies of the tax system in the U.S. resulted in consistent (inconsistent) findings, supporting the existence of tax-motivated shifts in the *timing* (*incidence*) of marriage and divorce (Sjoquist & Walker, 1995; Alm & Whittington, 1995). Both of these studies required the selection of a methodology for the computation of MTPs as one of their principal independent variables.

The Sjoquist and Walker study involved the hypothetical marriage of single males and females. Their "flow" variable approach ignored the number of divorces occurring over their study period (1948–1987). The Alm and Whittington study used actual panel data. Employing a "stock" variable approach, it included the impacts of MTPs on both marriages and divorces (1947–1988).

These studies spanned a period beginning with an increased female presence in the labor force (post-WWII), the broad use of newly developed birth control measures and their endorsement by the (U.S.) National Council of Churches (1961), a rise in feminism and the establishment of the National Organization for Women (NOW, 1963), the Equal Pay Act (1963), the Civil Rights Act (1964), gay rights activism, and the U.S. legalization of abortion by the Supreme Court (1973).

These events were not (specifically) included in the models developed by Sjoquist and Walker or Alm and Whittington. Instead, they were indirectly "captured" through the use of economic variables designed to incorporate the effects of these changes (e.g. single female earnings, female-to-male median income levels, female-to-male ratios, etc.) into their models. Both studies concluded that evidence of shifts in the *timing* of marriage were present and affected by changing MTPs.

A Single, Event-Based Study

Gelardi (1996), in his study of amendments to tax laws affecting marital decisions in Canada and England/Wales (E&W), also found evidence to support the position that tax law changes caused strategic shifts in the *timing* of marriage. Unlike U.S.-based studies, his conclusions were not based on the hypothetical "divorcing" or "marrying" of taxpayers or the calculation of MTPs. Instead, he studied the number of marriages before and after tax law changes (1985–1986 in Canada and 1967-8–1968-9 in E&W) designed to eliminate the tax benefits associated with fiscal year-end (December for Canada and March for E&W) marriages.

For E&W (1960–1991), the percentage of annual marriages occurring in March were highest for the 1967–1968 (fiscal) tax year. The sharpest decline in March marriages occurred during the 1969–1970 (fiscal) tax year. Similarly, for Canada (1950–1991), the percentage of annual marriages occurring in December were highest (lowest) for the 1985 (1988) tax year. Therefore, Gelardi's findings, that the *timing* of marital decisions can be influenced by relevant changes in tax law, is consistent with the U.S. findings developed by both Sjoquist and Walker and Alm and Whittington.

Appendix Table 8-4 summarizes contemporary studies of the marriage tax by study type, country and data/period studies, and the conclusions reached. Like Gelardi, the findings contained in this section do not involve the selection of a methodology for the calculation of MTPs and MTBs. Unlike Gelardi, this section involves the use of recent historical data for both marriages and divorces, in the previously unexplored form of the M-T-D ratio. Like Alm and Whittington, both marriage and divorce are examined, but in their net marriages form. The inclusion of both marriage and divorce "stock" variables led Alm and Whittington to conclude that the MTP contributed (significantly) to the *incidence* of marriage. These studies yield similar results for the findings developed in this section.

Comparisons of U.S. and Canadian M-T-Ds

A U.S.-Canada extension of the effects of differing tax policies of M-T-D ratios is of interest for several reasons. First, the Canadian "mosaic" and the U.S. "melting pot", while differing in their social and economic policies, both originated from British historical roots.

Second, both countries have had dramatic, comprehensive tax change packages in the 1980s (i.e. 1981 and 1986 in the U.S. and 1987 in Canada).

Third, Canada and the United States both employ a progressive personal income tax rate structure. Progressive tax rates are (generally) perceived to be fair and equitable, due to their compatibility with the *ability to pay* principle of taxation. Economic stability is also enhanced. Tax rate progressivity automatically shifts individuals to lower (higher) tax brackets during periods of economic downturn (expansion). However, problems, in the form of MTPs and MTBs, persist under a system of personal income tax rate progressivity where the family remains the unit of taxation.

Fourth, the personal income tax unit in Canada is the *individual*, while the U.S., providing for both joint and separate filing, taxes personal income at the *family unit* level. Canada provides for tax credits at a flat rate of 17%, based on family size (1988–). The United States provides non-itemizer taxpayers with a standard tax deduction, based on marital status and differing, progressive tax rates (see Appendix 2A and Chapter 6). Taxpayers able to itemize are subject to personal exemption phase-outs and itemized deduction limitations, having a "flattening" effect similar to the Canadian flat rate of 17%, but imposed at relatively high levels of income.

Finally, as summarized by MPL Communication, Inc. (1993, 4763)

    ... (a)mong the most important (trends in Canada) are the growth in common-law unions
    ...

Though a trend toward cohabitation may also persist in the United States, it might not have the same tax compliance-related implications. In Canada, the spousal or equivalent tax credit is reduced as income for the lesser earning spouse (including common law) rises.

Though Canadian measures are currently multiplied by a flat 17% (post-1987), a contemporary study by Rupert and Fischer (1995) concluded that U.S. taxpayer's perceptions of their marginal tax rates differ significantly from their actual rates.

The Calculation of M-T-D Ratios

Figure 8-2 provides for a graphic comparison of the M-T-D ratios for Canada and the United States. These measures consist of the number of marriages divided by the number of divorces, using equation (2), as follows:

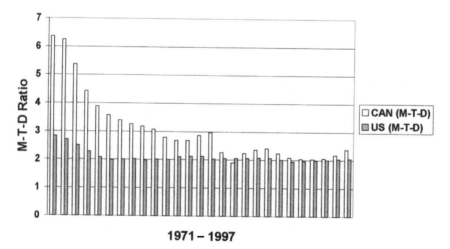

Fig. 8-2. A Comparison of Canadian (CAN) & U.S. (US) Marriage-to-Divorce (M-T-D) Ratios.

$$M\text{-}T\text{-}D \ Ratio = Marriages \div Divorces \qquad\qquad (8\text{-}2)$$

As Fig. 8-2 illustrates, the 1971–1997 period provides for relatively stable M-T-Ds for the United States, but produces a very sharp decline for Canada. The 1984–1987 tax years deviate from this downward trend and are particularly unstable for Canada. The post-1987 period is very stable for both Canada and the United States.

Appendix Table 8-1 summarizes the M-T-Ds used to generate Fig. 8-2. Also provided are Pearson product moment correlational measures for M-T-Ds and their components (N = 26). U.S. and Canadian measures of divorce and M-T-D ratios are highly (directly) correlated (at 84.7% and 89%, respectively) and statistically significant. U.S. and Canadian measures of marriage are not directly correlated, but are also not statistically significant. Finally, the correlation for both U.S. and Canadian measures of marriage and divorce are statistically significant, but the relation is a direct one for the U.S. and an indirect or inverse one for Canada. Canadian marriages have declined and Canadian divorces have nearly doubled over the 26 year period. U.S. and Canadian M-T-D ratios appear to have stabilized, at a 2-to-1 ratio, for over a decade (see Fig. 8-2).

Canadian Tax Policy Implications

Beginning with the 1993 tax year, Canadian tax law (ITA 252 (4)) required that the term "spouse" include situations where taxpayers have cohabited in a conjugal relationship for the preceding 12 month period. This affects the Canadian equivalent-to-spouse tax credit for dependents of individuals who are single, divorced, widowed, or separated. This tax credit can only be claimed if a spousal tax credit is not in effect (i.e. only one spouse or equivalent-to-spouse tax credit per household is allowed). The effect of this one spousal/equivalent-to-spouse tax credit per household is that a MTP may arise if two working individuals with one or more dependents (each) become cohabitants or get married. In this case, the Canadian equivalent-to-spouse tax credit for one or both spouses would be eliminated and replaced by one spousal tax credit, which is reduced by the earnings of the spouse with the lower income. The increased incidence of Canadian common-law unions combined with the Canadian decline in M-T-Ds might lead one to consider the possibility that these results may reflect evidence of taxpayer non-compliance with this relatively contemporary feature of Canadian tax law.

Table 8-1. U.S./Canadian Measures of Marriage, Divorce and Marriage-to-Divorce (M-T-D) Ratios (Marriage and Divorce Measures in Thousands – 1971–1997).

Tax Year	UNITED STATES			CANADA		
	Marriage	Divorce	M-T-D Ratio	Marriage	Divorce	M-T-D Ratio
1971	2,190	773	2.8331	191	30	6.3667
1972	2,282	845	2.7006	200	32	6.2500
1973	2,284	915	2.4962	199	37	5.3784
1974	2,230	977	2.2825	199	45	4.4222
1975	2,153	1,036	2.0782	198	51	3.8824
1976	2,155	1,083	1.9898	193	54	3.5741
1977	2,178	1,091	1.9963	187	55	3.4000
1978	2,282	1,130	2.0195	186	57	3.2632
1979	2,331	1,181	1.9738	188	59	3.1864
1980	2,390	1,189	2.0101	191	62	3.0806
1981	2,422	1,213	1.9967	190	68	2.7941
1982	2,456	1,170	2.0991	188	70	2.6857
1983	2,446	1,158	2.1123	185	69	2.6812
1984	2,477	1,169	2.1189	186	65	2.8615
1985	2,413	1,190	2.0277	184	62	2.9677
1986	2,407	1,178	2.0433	176	78	2.2564
1987	2,403	1,166	2.0609	182	96	1.8958
1988	2,396	1,167	2.0531	188	84	2.2381
1989	2,403	1,157	2.0769	191	81	2.3580
1990	2,443	1,182	2.0668	188	78	2.4103
1991	2,371	1,187	1.9975	172	77	2.2338
1992	2,362	1,215	1.9440	165	79	2.0886
1993	2,334	1,187	1.9663	159	78	2.0385
1994	2,362	1,191	1.9832	160	79	2.0253
1995	2,336	1,169	1.9983	160	78	2.0513
1996	2,344	1,150	2.0383	· 157	72	2.1806
1997	2,384	1,163	2.0499	159	67	2.3731

Pearson Product Moment Correlation (r):

$-28.4\%^a$	for $U.S._{Marriage} \times Canada_{Marriage}$
$84.7\%^b$	for $U.S._{Divorce} \times Canada_{Divorce}$
$89.0\%^a$	for $U.S._{M-T-D} \times Canada_{M-T-D}$
$64.7\%^a$	for $Divorce_{U.S.} \times Marriage_{U.S.}$
$-58.0\%^a$	for $Divorce_{Canada} \times Marriage_{Canada}$

[a] Not significant.
[b] Significant ($p < 0.001$).

Recommendations for Further Study

The Canadian/American tax treatment/policies for unmarried, co-habiting two-earner households/couples differs. In Canada (America), extended co-habitation does not (does) result in the circumvention of a MTP. These differences may provide the vehicle through which the impact of changing tax laws on the *incidence* of marriage might be isolated and quantified by academic researchers of these issues.

Furthermore, the (primarily) U.S.-based concerns over the demise of the "family unit", as operationally defined by declining (rising) marriage (divorce) rates, or the M-T-D ratio, may be the result of changes in the "form" (i.e. without benefit of a marriage license) and not the "substance" of the changing North American definition of the "family". Further examination of Canadian taxpayer compliance with ITA 252 (4) would provide insights into this question.

Preliminary Revenue Canada tax compliance-based investigations may begin with a simple matching of addresses for two taxpayers claiming to be single, but living at the same address. For researchers interested in this compliance issue, survey questionnaires directed primarily to the attention of tax professionals in both countries would draw on the insights available via preparer perceptions of the degree of cohabitation-based U.S. (Canadian) tax avoidance (evasion).

Finally, longitudinal comparisons of the ratio of annual tax returns filed to the annual number of households in each country might also be investigated and controlled for. For example, if the Canadian decline in the M-T-D ratio is a function of non-compliance with Canadian tax law, one would expect a decrease in the correlation between the number (and classification) of household-based tax returns *anticipated* and the number (and classification) of tax returns actually *filed* each year.

SUMMARY

As this monograph goes to press, the "Marriage Tax Penalty Reconciliation Act of 2000" (H. R. 4810) is proceeding through the U.S. House and Senate. This and other bills designed to reduce or eliminate the MTP have not been successful.

The four versions of this Bill, in aggregate, provide for the elimination of both rate and base effects, adjustments for the earned income tax credit (discussed in Chapter 7), and adjustment for the alternative minimum tax (discussed in Chapter 9). They were designed to reduce taxes paid only by married couples who both work. The decomposed impact of the "marriage tax" for all of these measures were quantified in Cataldo (1996).

Vice-President Gore, in his Presidential campaign, proposed an increase in the standard deduction (noted in Chapter 6), resulting in the reduction or elimination of the base effect component of the MTP for married couples. President Bush suggested the restoration of the special deduction of up to $3,000 (i.e. 10% of $30,000) for two-earner couples (discussed in Appendix 3H). At a minimum, we anticipate passage of the relatively low cost Bush proposal.

Contemporary low unemployment rates in the U.S. and changing demographic trends, combined with anticipated U.S. budget surpluses, suggests that tax policy may increasingly favor families. Appendix 8M provides some historical and contemporary international evidence on family taxation.

PART 3

CAPITAL GAIN/LOSS TAXATION

CHAPTER 9

THE ALTERNATIVE MINIMUM TAX

The individual alternative minimum tax (AMT) is a complicated tax that currently affects relatively few taxpayers (700,000 in 1997) and raises relatively little revenue ($4.5 billion in 1997). By 2007, however, the number of AMT taxpayers will reach 9 million and their AMT liability will reach $21 billion. The reason . . . personal exemption and rate bracket widths . . . are indexed for inflation, whereas . . . AMT exemption . . . (is) not (Harvey & Tempalski, 1997, 453).

INTRODUCTION

Historically, the alternative tax(es) and preferential treatment for long-term capital gains and losses were inter-related. We have attempted, in our coverage, to separate these tax policies into two chapters. This chapter's primary focus is on the contemporary alternative minimum tax and its predecessors.

The first alternative tax, providing for the preferential treatment of long-term capital gains, was in effect for the 1922–1933 tax years. The second alternative tax had similar objectives, was modified frequently, and was effective for the 1938–1978 tax years. These historical "alternative tax" (AT) measures were designed to *reduce* otherwise ordinary tax rates. They were very different from the contemporary alternative minimum tax (AMT; 1979–), which originated and evolved from the "add-on" minimum tax on tax preferences (MINTAX; 1970–1982). At the same time that a 10% surcharge was imposed for the 1968 tax year.

Because this surcharge was only in effect for the last nine months of the 1968 tax year (i.e. the tax was first imposed on the number of days occurring after April 1, 1968), it effectively took the form of a 7.5% surcharge (IRS SOI, 1969, 1). The objective was to raise Federal tax revenues. For the 1968 tax year, the surcharge raised $5.2 billion in Federal tax revenues. Approximately 49.2 million of the 61.9 million individual Federal income tax returns filed for the 1968 tax year were affected by the surcharge (IRS SOI, 1968, 1–2).

The Canadian "minimum tax" (MT; 1986–) is also discussed in this chapter. This relatively new feature of the Canadian system of individual federal taxation appears to have been patterned after the U.S. AMT.

The chapter that follows, Chapter 10, focuses on capital gains and losses, per se. Chapter 10 also expands on the policy issues relating to the preferential treatment of long-term capital gains.

THE FIRST ALTERNATIVE TAX – 1922–1933

Under the Revenue Act of 1921 and subsequent acts, capital net gains as defined in the revenue acts from the sale of assets held more than 2 years may, at the option of the taxpayer, be reported separately and taxed at $12\frac{1}{2}\%$ in lieu of the normal tax and surtax rates otherwise applicable (IRS SOI, 1931, 7).

The first alternative tax (AT) provided for the preferential treatment of net long-term capital gains (LTCGs). KixMiller and Baar (1921, 139–140) explained the underlying rationale:

Because the gain realized upon the sale of a capital asset may have accrued over several years, a special treatment is adopted . . . only when realized after December 31, 1921 . . . the tax . . . shall be $12\frac{1}{2}\%$ thereof, in lieu of any other normal tax or surtax . . . This means it is not advantageous unless the net income is large, over $30,000 at least.

Therefore, an "alternative" to normal tax and surtax rates was available in the form of a preferentially lower tax rate or ceiling of $12\frac{1}{2}\%$. Since only 6.28% (IRS SOI, 1921, 2) of the U.S. population filed tax returns for the 1921 tax year, and slightly more than 40% (IRS SOI, 1921, 4) of the tax returns filed for the 1921 tax year were at or below the $25,000 income class, only the very highest of the high-income U.S. taxpayers were in a position to take advantage of this preferential rate of $12\frac{1}{2}\%$. This preferential treatment was not referred to as an "alternative" tax at the time.

To illustrate the significance of the tax savings made available through this *alternative*, the 1922 IRS SOI (13) provides a perspective on historical surtax rates:

For 1916 the highest rate of surtax was 13%. For 1917 it was 63% plus a war excess profits tax on business income, and for 1918–1921, inclusive, it was 65%. For 1922 it was 50% on net income other than gain from the sale of capital assets held for more than two years, on which the rate was $12\frac{1}{2}\%$.

Appendix Table 9-1 provides a summary of selected measures for LTCG exempted from the *normal* tax, but potentially subject to the *surtax* rates (discussed above). Similarly, long-term capital losses (LTCLs) were not deductible in calculating the *normal* tax during this period. However, after the

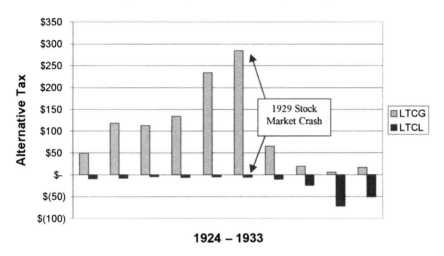

Fig. 9-1. The First *Alternative* Tax (000 omitted).

1923 tax year, LTCLs were provided comparable treatment, in the form of the same $12\frac{1}{2}\%$ tax rate afforded LTCGs (IRS SOI, 1924, 7). These measures are also provided in Appendix Table 9-1.

Figure 9-1 graphically illustrates both LTCGs and LTCLs (at the $12\frac{1}{2}\%$ tax *credit* rate) for the 1924–1933 tax years. It also shows the impact of the 1929 stock market crash for the years that followed. The two-year holding period for preferential *alternative* tax rate treatment is also apparent, as LTCLs peaked for the 1932 tax year.

For the tax years following 1933, the alternative $12\frac{1}{2}\%$ tax on LTCGs was repealed and replaced with varying percentages, based on holding periods (discussed in Chapter 10).

THE SECOND ALTERNATIVE TAX – 1938–1978

The major changes provided by the Revenue Act of 1938 . . . the application of an alternative tax in the case of returns with net long-term capital gains and losses . . . (IRS SOI, 1938, 3).

The second alternative tax (AT) also targeted taxpayers with LTCGs and LTCLs. As provided by the Revenue Act of 1938, the second AT also had the

potential to result in a reduction in total tax (normal tax *plus* surtax) for
taxpayers with net LTCG.

> The alternative tax is imposed on returns with net long-term capital gain if such alternative
> tax is less than the sum of the normal tax and surtax computed on net income including net
> long-term capital gain. The alternative tax is imposed on returns with net long-term capital
> loss if such alternative tax is greater than the sum of the normal tax and surtax computed
> on net income after deducting net long-term capital loss (IRS SOI, 1938, 10).
>
> Net long-term capital gains are included with other income subject to normal tax and
> surtax rates or are segregated and taxed at 30%, whichever method results in the lesser total
> tax. Net long-term capital losses are deducted from other income and a tax is computed, or
> 30% of such losses is credited against the tax computed on net income before deducting the
> net loss, whichever method results in the greater tax (IRS SOI, 1938, 3).

This second AT coincided with (1) revisions in the treatment of net short-term
capital gain (STCG)[90] and LTCG and current loss tax rates, and (2) holding
periods for short-term and long-term status. Equations (9-1a) and (9-1b)
summarize the mechanics of the AT as it related to net LTCG (and tax
reductions) or net LTCL (and tax increases), respectively:

$$AT = lesser\ of: \tag{9-1a}$$
$$Regular\ Tax\ (including\ net\ LTCG)$$
$$or$$
$$Regular\ Tax\ (excluding\ net\ LTCG) + (30\% \times LTCG)$$

$$AT = greater\ of: \tag{9-1b}$$
$$Regular\ Tax\ (reduced\ by\ net\ LTCL)$$
$$or$$
$$Regular\ Tax\ (excluding\ net\ LTCL) - (30\% \times LTCL)$$

Appendix Table 9-2 contains a summary of the number and percentage of tax
returns filed and affected by the AT for the 1938 and subsequent tax years. At
no time, throughout the history of this second AT (1938–1978), did it affect
even 1% of the tax returns.

Tax savings (or net tax savings) resulting from the use of this second AT
were not provided by the 1938 IRS SOI, but separate AT measures were
provided as they related to net LTCG ($150.6 million) (IRS SOI, 1938, 88) and
net LTCL ($128.9 million) (IRS SOI, 1938, 90). Therefore, the means of
crudely approximating the early AT net tax savings, by individual taxpayer net
income class, are available for researchers interested in this component of the
historical perspective of capital gains taxation.

The Revenue Act of 1941, affecting individual tax returns with a tax year
beginning after December 31, 1940, included the implementation of an
"optional tax", also designated as an AT. This provision applied to individuals

with certain gross income of $3,000 or less (provided in Supplement T of the IRC), who elected to file the Form 1040A. Like the existing AT, this AT was in lieu of the normal tax and surtax (IRS SOI, 1941, 3).

The AT was modified by the Revenue Act of 1942 (for 1942 and later tax years), and continued to provide a potential vehicle for tax reduction. This revised AT applied to returns with an excess of net LTCG over short-term capital loss (STCL), but only if the AT was less than the combined normal tax and surtax on taxpayer net income, which included the net gain from the sale/ exchange of capital assets (IRS SOI, 1942, 9 and 78).

A capital loss was allowed for the lesser of (1) gains plus other net income, or (2) $1,000. Formerly, deductions for the STCL were permitted only to the extent of STCG, and LTCL was permitted only to the extent of LTCG and other income, including net STCG. The rate applied to LTCG was changed from 30% of net LTCG to 50% of the excess of net LTCG over net STCL (IRS SOI, 1942, 4). This modification is reflected in Eq. (9-2), as follows:

$$\text{AT} = lesser\ of: \tag{9-2}$$
$$Regular\ Tax\ (including\ net\ LTCG)$$
$$\text{or}$$
$$Regular\ Tax\ (excluding\ net\ LTCG) + (50\% \times (net\ LTCG - net\ STCL))$$

For the 1945 tax year, tax returns containing AT continued to be "long-form" returns, but applied only to taxpayers with surtax net income (see Chapter 3) above $16,000 (IRS SOI, 1945, 10). The "optional tax", under Supplement T of the Internal Revenue Code, also remained available (IRS SOI, 1945, 8), and was adjusted to reflect additional post-1945 tax year exemptions and additional reductions in normal and surtax rates (IRS SOI, 1946, 7).

For the 1946 and 1947 tax years, the AT was applicable only in cases where surtax net income was greater than or equal to $18,000 (IRS SOI, 1946, 9). Beginning with the 1948 tax year, this threshold was increased to $22,000 for separate returns and $44,000 for joint returns filed by married taxpayers (IRS SOI, 1948, 32–33). These measures were reduced to $20,000 ($40,000), beginning with the 1950 tax year (IRS SOI, 1950, 21), and further reduced to $16,000 ($32,000) for the 1951 tax year (IRS SOI, 1951, 13). Beginning with the 1952 tax year, the computation of the AT changed, and is represented by Eq. (9-3), as follows:

$$\text{AT} = [Regular\ Tax\ (excluding\ net\ LTCG) \tag{9-3}$$
$$- (50\% \times (net\ LTCG - net\ STCL))]$$
$$+ [26\% \times (net\ LTCG - net\ STCL)]$$

For the 1952 tax year the AT was applicable only in cases with surtax net income greater than or equal to $14,000 for separate returns, $28,000 for joint returns, and $20,000 for the returns of heads of households (IRS SOI, 1952, 14). Beginning with the 1953 tax year, the AT, which limited the effective income tax rate on excess net LTCG over net STCL to 25%, proved advantageous when taxable income reached $18,000 on a separate return, $36,000 on a joint return, or $24,000 for a taxpayer filing a return under the head of household status (IRS SOI, 1955, 11 and 1958, 23), and is represented by Eq. (9-4), as follows:

$$AT = lesser\ of: \qquad\qquad (9\text{-}4)$$
$$Regular\ Tax$$
$$or$$
$$[Regular\ Tax\ (excluding\ (net\ LTCG - net\ STCL))]$$
$$+ [25\% \times (net\ LTCG - net\ STCL)]$$

Throughout this entire period, the second AT resulted in tax reductions, when compared to non-preferentially treated tax rates.

Beginning with the 1964 tax year, the IRS SOI (1964, 122) explained:

> (p)roviding there were some capital gains, the alternative computation of tax was advantageous if taxable income other than capital gains exceeded $40,000 on joint returns and returns of surviving spouse, $32,000 on returns of heads of households, or $20,000 on separate returns of other persons. These were the points at which the marginal combined normal tax and surtax rates on the different rate schedules exceeded 50%.

Beginning with the 1965 tax year, the AT proved advantageous when taxable income reached $26,000 on a separate return, $52,000 on a joint return, or $38,000 for a head of household (IRS SOI, 1965, 52; 1968, 176; and 1969, 120).

For the 1969 tax year, a 13% ($10 billion) increase in Federal tax revenues resulted from: (1) a 3% increase in the number of individual Federal income tax returns filed, (2) an increase of 9% in taxpayer AGI, and (3) the extension of the tax surcharge on individual income before tax credits, which covered the entire 1969 calendar and tax year (IRS SOI, 1969, 1). The impact of the tax surcharge was explained in the 1969 IRS SOI (123):

> After the taxpayer computed his regular tax liability, he then had to increase this amount by 10% of the tax reduced by any retirement income credit. This addition to tax constituted the tax surcharge imposed for the period January 1, 1969, through December 31, 1969.

The 1969 tax year also provided for an alternative method of tax computation for taxpayers with large amounts of capital gains. The taxable half of net long-term capital gains (in excess of net short-term capital losses) was taxed at a 50% rate (IRS SOI, 1969, 119–120). Of the 64.2 million individual Federal income tax returns filed for the 1969 tax year, only 0.1 million found the AT advantageous. The AT established an individual Federal income tax rate at a 25% maximum for net long-term capital gains.

The 1970 tax year saw: (1) the introduction of a tax on specified "tax preferences", (2) the imposition of higher rates on capital gains, and (3) a limitation on the capital loss deduction (IRS SOI, 1970, iv). It was at this time that the AT began to take its contemporary form (IRS SOI, 1970, 149).

> Under prior law, a taxpayer with large amounts of taxable income could elect an alternative tax computation which limited to 50% the tax on the taxable half of excess net long-term capital gain over net short-term capital loss. The Tax Reform Act of 1969 (TRA69) continued this rate on the first $25,000 of such income ($12,500 in the case of married persons filing separately), but applied a 59% rate for 1970 to amounts in excess of this base.

"For 1971, this rate was raised to 65%" (IRS SOI, 1971, 123). "For 1973, as for 1972, the excess could only be taxed at the regular rates" (IRS SOI, 1973, 71). The 25% AT, applying to the first $50,000 of net capital gain, was eliminated for 1979 and later tax years:

> (t)he alternative tax computation for taxpayers with long-term capital gains and marginal tax rates above 50% was abolished; on the other hand, the 60% exclusion of long-term capital gains, introduced for sales of capital assets after October 31, 1978, became effective for the full tax year on returns filed for 1979 . . . (IRS SOI, 1979, viii).

THE ADDITIONAL (MINIMUM) TAX FOR TAX PREFERENCES – 1970–1982

> The purpose of the additional tax was to make possible the taxation, to some extent, of items previously accorded special treatment (IRS SOI, 1970, 150).
> The "add-on" minimum tax was introduced in 1969 in an attempt to obtain some tax contribution from wealthy people who had previously escaped income taxation on all or most of their income. The tax was levied at a 10% rate on a selected list of "preference" incomes to the extent that they exceeded $30,000 plus the regular income tax (Pechman, 1987, 128).

The 10% "add-on tax", "additional tax for tax preferences", or "minimum tax" (MINTAX) was introduced by TRA69. A *direct* predecessor to the contemporary alternative minimum tax (AMT), the computation of the MINTAX was detailed on Form 4625, as follows:

- If totaled tax preferences were ≥ $15,000, a Form 4625 was required,
- The exclusion was $30,000 ($15,000 for married taxpayers, filing separately),
- Tax preferences were reduced by income tax after credits and adjusted by a "no benefits exclusion" (i.e. tax preferences from which no tax benefit was derived),

- A "tentative tax on tax preferences" was computed at 10%,
- Was reduced by applicable 1970 "net operating loss carryover", and
- There was further reduction for any unused retirement income credit.

The MINTAX was imposed on approximately 18 to 19 thousand of the 75.3 thousand individual Federal income tax returns with tax preferences (IRS SOI, 1970, 150). This compares to approximately 74.3 million individual Federal income tax returns filed for the 1970 tax year.

Appendix Table 9-3 provides a summary of the percent of individual federal income tax returns affected by the MINTAX. As was the case with the alternative tax (discussed above), the MINTAX did not even affect even 1% of the U.S. tax returns filed for any single tax year.

For the 1971 tax year, there was a 46% increase in net capital gains. This increase followed two consecutive years (1969 and 1970) of decreases. The MINTAX affected approximately 24 thousand individual federal income tax returns (see Appendix Table 9-3). Approximately four thousand of these showed no tax under the regular individual federal income tax computation. This compares to approximately 74.6 million individual federal income tax returns filed for the 1971 tax year.

Related measures increased slightly for the 1972 and 1973 tax years. The MINTAX affected approximately 27 (1972) and 26 (1973) thousand individual Federal income tax returns and approximately four (1972) and six (1973) thousand, respectively, showed no tax under the regular individual federal income tax computation (see Appendix Table 9-3). This compares to approximately 77.6 (1972) and 80.7 (1973) million tax returns filed for the 1972 tax year.

Beginning with the 1976 tax year, TRA76: (1) expanded the definition of "tax preferences" (including the major tax preference item associated with itemized deductions, where in excess of 60% of AGI), (2) increased the MINTAX from 10% to 15%, and (3) reduced the deductions previously available for the reduction of the MINTAX. These modifications resulted in a significant increase in the percentage of taxpayers affected by the MINTAX (see Appendix Table 9-3).

The establishment of the alternative minimum tax (AMT), a result of TRA78 and beginning with the 1979 tax year, resulted in a significant reduction in the applicability of the MINTAX (see Appendix Table 9-3).

ERTA81 provided for tax law changes to be phased-in over several years. It (initially) affected the computation of the MINTAX, which included a new tax preference item – depreciation in excess of straight-line amounts – under the accelerated cost recovery system (ACRS), at the MINTAX rate of 15%.

The MINTAX was abolished after the 1982 tax year. The revised AMT covered many of the tax preference items were previously covered by the MINTAX, and the AMT rate was raised to 20% (IRS SOI, 1983, 1).

THE CONTEMPORARY ALTERNATIVE MINIMUM TAX (1979–)

The alternative minimum tax's (AMT) purpose is to ensure that all taxpayers share the tax burden fairly. It prevents a taxpayer with substantial income from avoiding significant tax liability (RIA, 1995, 490).

TRA76 required the annual publication of "information on the amount of tax paid by individual taxpayers with high total incomes", in addition to "the number of such individuals . . . who owe no Federal income tax" (IRS SOI, 1975, 3). The law specified the use of three new income concepts:

(1) AGI *plus* tax preferences,
(2) AGI *less* investment interest, and
(3) AGI *plus* tax preferences *less* investment interest.

As the IRS SOI (1976, 82) notes:

The Tax Reform Act of 1976 brought about a nearly sevenfold increase in the additional tax by expanding the definition of tax preferences, by reducing the deductions from total tax preferences in arriving at tax preferences subject to tax, and by increasing the tax rate on tax preferences subject to tax from 10 to 15%.

For the 1977 tax year, a Form 4625 was required only for taxpayers with $10,000 ($5,000 if married, filing separately) or more of total tax preferences

TRA78 included the establishment of the *progressive* alternative minimum tax (AMT).[91] This tax was generally computed by adding to taxable income any excluded LTCGs plus "excess" itemized deductions (those exceeding 60% of adjusted gross income), subtracting $20,000, and subjecting the remainder to a graduated tax ranging from 10 to 25%. The alternative minimum tax was then reduced by other income taxes (IRS SOI, 1979, 60).

With a maximum rate of 25%, the AMT evolved from the separation of the additional tax for tax preferences into: (1) the AMT, applying to itemized deductions and capital gains tax preferences, and (2) the previously established *flat* 15% MINTAX for all other tax preferences. More than 133,000 taxpayers, otherwise not subject to tax under normal tax computation methods, were required to pay the AMT (IRS SOI, 1979, 59). For the 1980 tax year, the only tax credit that could be used to completely offset the AMT was the foreign tax credit (IRS SOI, 1980, 8).

For the 1979 tax year there was both: (1) a MINTAX (at a flat rate of 15%) and (2) an AMT (at progressive rates, with a maximum of 25%). The taxpayer was required to pay the greater of the tax computed under: (1) otherwise allowable tax computation methods, or (2) the AMT method (see Appendix Table 9-3). The result was an increase from an average additional tax of $3,057 for 1978 to $5,293 for 1979 (IRS SOI, 1979, 61).

Appendix Table 9-4 provides summary measures for the contemporary AMT. As was the case with its predecessors, the AMT has not even affected one% of the taxpayers for any single tax year.

ERTA81 modified the AMT. Its major provisions were intended to increase savings and investment (IRS SOI, 1981, 6). The aggregate AMT increased by approximately 50% (from $0.85 billion to $1.26 billion), while the maximum rate applied for the purposes of the LTCG related component of the AMT was reduced from 25% to 20%, for transactions entered into after June 9, 1981.

For the 1982 tax year, the AMT consisted of three progressive brackets (0, 10 and 20%). Beginning with the 1982 tax year, the maximum tax on "personal service" income was repealed. Beginning with the 1983 tax year, the MINTAX was abolished and the AMT was revised by TEFRA82. The AMT was raised to a flat 20% of alternative minimum taxable income (AMTI).

TRA86 further modified the AMT. Beginning with the 1987 tax year, the starting point for calculating AMTI shifted from AGI to taxable income (TI). The personal exemption amount was no longer available as a deduction for (AMTI) calculation. In addition, the flat AMT rate was increased from 20% to 21%. This rate was increased to 24%, beginning with the 1991 tax year, and increased, again, to a range of 26 to 28%, beginning with the 1993 tax year.

TRA97 modified the AMT for capital asset sales and exchanges occurring after May 6, 1997. Generally, the 10% and 20% capital gains rates will apply for these transactions. For tax years beginning after 2000, the 8% and 18% capital gains rates applicable to property with a five-year depreciable life will also be applied for AMT purposes (RIA, 1998, 549).

THE CANADIAN MINIMUM TAX (1986–)

For many years, a common complaint about the Canadian tax system was that an individual could have a six figure income and still pay only minimal income taxes. While such cases involve no more than taking full advantage of the various provisions in the Act that allow individuals to reduce their taxes payable, there was a strong public feeling that allowing wealthy individuals with high levels of economic income to pay little or no taxes is not an equitable situation. To deal with this, an alternative minimum tax was introduced in 1986 (Byrd, Chen & Jacobs, 1994–1995, 341).

The Canadian minimum tax (MT) began with the 1986 tax year. Although we have not discussed it in detail, its goals and objectives are comparable to those for the contemporary U.S. AMT. Appendix Table 9-5 provides summary measures for the contemporary Canadian MT. As was the case with the U.S. counterpart and predecessors, the MT has not even affected 1% of taxpayers for any single tax year.

A COMPARISON OF THE U.S. AMT AND CANADIAN MT

Figure 9-2 provides a timeline of the evolution of U.S. MINTAX (1970–1982), U.S. AMT (1979–), and Canadian MT (1986–). Similarly, the percentage of taxpayers affected by the contemporary U.S. AMT and the Canadian MT (1986–1992) is provided in Fig. 9-3.

These figures serve to illustrate the flow of tax policy ideas, and their implementation, between the U.S. and Canada. Just as the Canadian tax system first embraced the notion of tax credits for dependent children (see Chapter 6), the U.S. appears to have led with the notion of "minimum taxation".

UNITED STATES (1970–)

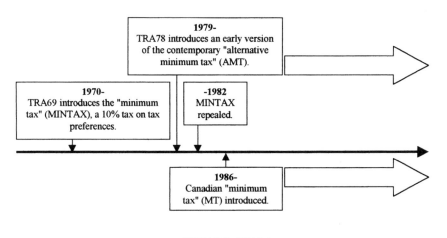

CANADA (1986–)

Fig. 9-2. A Timeline Comparison of Effective Dates for the U.S. MINTAX (1970–1982) and AMT (1979–) and the Canadian MT (1986–).

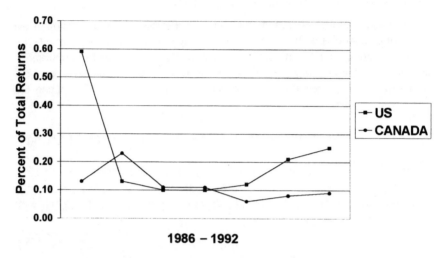

Fig. 9-3. A Comparison of the Contemporary U.S./Canadian AMT/MT.

SUMMARY

This chapter has introduced five separate evolutionary measures relating to "alternatives" to regular tax rates. Four are from the U.S., while one relates to the Canadian tax system.

Of the four U.S. "alternatives" the first two dealt with a *reduction* in the otherwise applicable tax rates for long-term capital gains. These "alternative tax" (AT) calculations were introduced for the 1922–1933 tax years (a period when only very high-income persons filed individual federal income tax returns), repealed, and reintroduced – with numerous modifications – for the 1938–1978 tax years. Even though a relatively small percentage of the U.S. tax-filing population were affected by these alternative taxes (for examples, see Appendix Table 9-2), these provisions were passed into law based on concerns over the equitable treatment of long-term capital gains.

The latter two relatively contemporary U.S. "alternative tax" calculations also affect only a tiny percentage of high-income taxpayers. The MINTAX (1970–1982) and the AMT (1979–) serve the same purpose of equity. The pattern of behavior in the U.S. individual federal income tax system is to pursue the perception, if not also the reality, of equity. The contemporary AMT also serves this purpose. However, the contemporary AMT is designed to prevent high-income taxpayers from completely escaping taxation through aggressive, tax preference-based tax planning strategies. The Canadian MT (1986–) was designed to serve the same purpose.

The historical evolution of these methods of addressing "tax preference" items is long, and we expect some form of the AMT to continue to apply to U.S. individual taxpayers. In the absence of a move toward a flat tax, a national sales tax, or extended U.S. budget surpluses, we do not expect that it will be politically possible to completely eliminate the AMT. However, it is equally likely that a significant increase in the percentage of the population adversely affected by the AMT will be politically infeasible.

We expect the U.S. AMT to modified, though not eliminated, within the next decade. This modification will probably take a form that continues to provide for an AMT that affects only a very small percentage of very high-income taxpayers. Though periodic, statutory increases in the AMT exemption are a possibility, inflation indexing of the AMT exemption would provide for a self-adjusting mechanism. The tax cut proposal package suggested by President Bush does not provide for inflation indexing of the AMT exemption measure.

CHAPTER 10
CAPITAL GAINS AND LOSSES

Rising incomes, increasing wealth, and the desire of taxpayers to avoid taxes, have combined with a substantial extension of the area eligible for capital gains rates to produce a rapid growth in capital gains. The yield from the capital gains tax has grown approximately 10% a year since the end of World War II, while the overall yield of income taxes has grown at half that rate (David, 1968, 229).

The ... economist's comprehensive definition of income for household units ... includes ... capital gains accrued during the year (whether realized or not) (Pechman & Okner, 1974, 2).

In 1974, 1975 and 1976 the House Ways and Means Committee and the Senate Finance Committee have considered formulas that would permit taxpayers to exclude from taxation a larger proportion of each nominal capital gain the longer each asset has been held. Unfortunately, this argument confuses *absolute* with *proportional* gains (Brinner, 1976, 126).

Few countries tax capital gains (Tanzi, 1976, 221).

... capital gains are heavily concentrated in the higher income classes ... (Pechman, 1987, 119).

... a capital gains tax cut ... is more beneficial to mature firms than new startups ... contradicts the widely held view that a capital gains tax cut would be a well-targeted approach for encouraging new firm capital formation (McGee, 1998, 653).

Until 1972, Canadian tax legislation did not levy any income tax on capital gains (Byrd, Chen & Jacobs, 1997–1998, 257).

INTRODUCTION

Ordinary income is taxed according to the appropriate rate established in the IRC schedules for single, married, and head-of-household taxpayers. *Capital gains* are either fully or partially included in ordinary income, or may be taxed at some fixed alternative rate rather than that otherwise specified in the Internal Revenue Code schedules. Therefore, in the absence of other over-riding concerns, the preferential or tax-favored treatment of long-term capital gains represents a violation of the concept of both horizontal and vertical equity (see Chapter 5). According to David 1968 (54):

On the assumption that capital gains are income, preferential capital gains taxation robs the tax structure of both (horizontal and vertical) equity. The (preferential treatment of long-term capital gains) reduces the progression of legal bracket rates . . . (and increases) . . . the ability of high-income persons to use their wealth for tax avoidance through capital gains provisions at the same time that they are able to take advantage of their skills and knowledge of the asset market to maximize return.

Alternatively, Pechman (1971, 27) classifies the "preferential treatment of capital gains" in a section of his book on U.S. tax policy devoted to "increasing saving and investment incentive" and as a "feature specifically intended to promote saving and investment".

CAPITAL GAINS TAXATION – THE U.S., CANADA, AND OTHER NATIONS

The U.S. taxation of gains on capital assets began with the Revenue Act of 1921. Canada did not tax capital gains prior to the 1972 calendar year.

Figure 10-1 graphically illustrates the percentage of individual income tax returns with capital gains and losses for both the U.S. and Canada (1972–1996). U.S. measures have been approximately double those for Canada. Because the Canadian *Taxation Statistics* publications provide separate annual measures for capital gains and losses by category, it was also

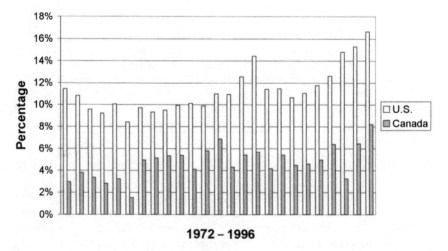

1972 – 1996

Fig. 10-1. A Comparison of U.S. and Canadian Taxpayers with Capital Gains and Losses.

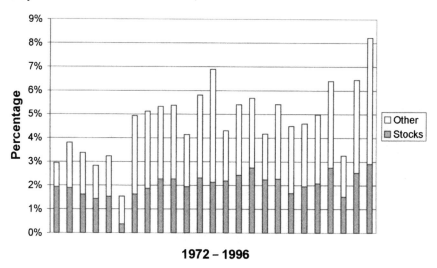

Fig. 10-2. Canadian Taxpayers with Capital Gains and Losses.

possible to produce Fig. 10-2, illustrating stock related gains and losses relative to those arising from other capital assets.

How do the U.S. and Canada compare to other nations? The data summarized in Appendix Table 10-1 was used to produce Fig. 10-3, graphically representing those countries (as of January 1, 1991) with and without capital gains taxation. The U.S. and Canada are in the majority. As of the early 1990s, approximately 65% of the 100 countries listed in Appendix Table 10-1 taxed capital gains.

Appendix Table 10-2 summarizes selected descriptive measures for the entire U.S. history of capital gains taxation (1922–1996). Appendix Table 10-3 summarizes similar measures, but for Canada (1972–1996). Appendix Table 10-4 provides Canadian measures as well, but only for stocks (1972–1996).

THE U.S. SYSTEM

As explained in KixMiller and Baar's *1922 United States Income and War Tax Guide* (139–143):

> Because the gain realized upon the sale of a capital asset may have accrued over several years, a special treatment is adopted for such gain, but only when realized after December 31, 1921.
>
> A profit accrued before March 1, 1913, that is, before income was taxable, is not taxable even though it was realized after that date. The fair market value as of March 1, 1913, when

Fig. 10-3. Countries with (and without) Capital Gains Taxation (shaded) – as of January 1, 1991.

greater than cost, is deducted from the selling price of property purchased before that date to determine the taxable profit.

The Return must state how the fair value as of March 1, 1913, is determined.

U.S. Tax Law distinguishes between receipts and costs for three classes of potentially taxable income:

(1) *Ordinary income*, taxed in full after exemptions and deductions,
(2) *Deductions*, fully deductible only after reduced by exclusions, and
(3) *Capital gains*, partially included in the tax base.

Figure 10-4 provides a basic overview of contemporary U.S. capital gains/ losses taxation.

The special or "preferential" treatment afforded to long-term capital gains (LTCG) arises from the desire to motivate taxpayers to engage in certain behaviors.

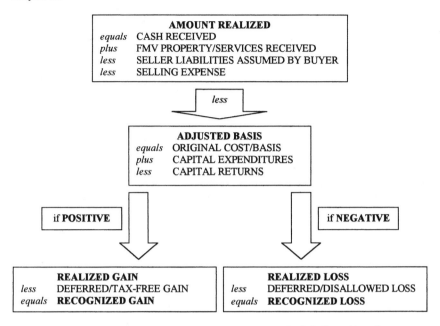

Fig. 10-4. A Basic Overview of Contemporary U.S. Gain/Loss Taxation.

The first objective for the preferential treatment of LTCG is to provide an economic incentive for greater *risk-taking*. Generally, a rational investor is willing to undertake a greater risk only if afforded the opportunity to generate a higher return on his/her investment. A lower tax rate on any capital gains or the exclusion of a portion of the gain to which ordinary income tax rates are applied increases the after-tax return to the risk-taking taxpayer. Therefore, the taxpayer who is willing to defer the consumption of his/her current economic resources, and is willing invest these resources in some risky economic venture, is treated relatively favorably.

The second objective is to encourage real investment, particularly in the form of new (and, presumably, risky) small (or even large) *business formation*. New businesses require economic resource inputs (e.g. facilities, inventory, personnel, and so on). Therefore, the formation of new businesses presumably seeking to produce the proverbial better mousetrap, provides the foundation for economic expansion and a broadening of the existing tax base. This occurs in the form of demand for assets, services (from other businesses), and increased employment opportunities. The taxpayer is contributing to the U.S. economy by undertaking profit-motivated risks and is rewarded, if the venture is successful, via the tax-favored treatment of any capital gains that might result.

Of course, all ventures of this sort are not successful. The risk-taking taxpayer may experience capital losses, which are not afforded preferential treatment. In fact, the deductibility of net losses is restricted to $3,000 per year, per federal income tax return. Clearly, the U.S. Government does not want to subsidize risky ventures above this annual ceiling.

The presence of the risk of loss in any risk-taking venture leads to the third objective of preferential LTCG treatment. Here, the objective cited is to provide for the averaging of volatile income levels that might results from capital gains in one tax year and capital losses in another.

To reduce taxpayer motivation toward highly speculative behaviors, a fourth objective is pursued. The imposition of *minimum holding periods*, for those taxpayers wishing to achieve LTCG treatment or status, in addition to the annual deductible net loss limitation, is intended to further dissuade speculation. The objective is to provide for stability in capital markets.

Alternative approaches to capital gains taxation were examined by David (1968) over thirty years ago. He first classified the economic impact of capital gains taxation as consisting of (David, 1968, 2–5):

(1) Growth effects – Increased returns for businesses and households willing to postpone consumption increases the *demand* for investment, and businesses increase the *supply* of stock-appreciating securities by increasing their savings rates, and decreasing their dividend payout percentages so that corporate retained earnings will rise; real investment is financed by way of these internally financed corporate savings;

(2) Efficiency effects – Tax-induced changes in the relative before-tax and after-tax yields of alternative investments affect the allocation of economic resources among available alternatives, the demand for legal and accounting services for tax compliance and tax planning divert professional and managerial talent away from productivity enhancing efforts; and the supply of labor may be affected by the preferential treatment of lower capital gains tax rates; and

(3) Stabilization effects – *Nondiscretionary* responses include increases (decreases) in tax yields achieved during inflationary (deflationary) periods; *discretionary* responses involve the latitude provided the taxpayer (e.g. realization) in the timing of capital gain-related income tax payment.

The issues raised by David (1968, 5–8) remain relevant in the contemporary and prospective periods, as follows:

(1) Do capital gains represent income?

(2) Should capital gains be taxed annually, upon realization, or at some other time?
(3) Should capital gains be taxed at progressive rates?
(4) Should capital gains escape taxation through gift or bequest at death?
(5) Should capital losses be permitted as partial or full offsets to ordinary income?
(6) How does capital gains taxation affect risk and liquidity in capital markets?

(1) Does the Capital Gain Represent Income?

> If a business is going to continue to operate as a going concern, it will usually have to replace any capital assets that are sold. Since gains on the sale of capital assets often reflect nothing more than inflationary price increases, such gains cannot be distributed to the owners of the business as they must be used to finance the replacement of the assets sold (Byrd, Chen & Jacobs, 1994–1995, 215).

Some argue that capital gains cannot represent taxable income, since all or part of the capital gain represents inflation. David suggests the analogy of a tree and its fruit. The fruit represents the annual income, which is subject to taxation. Pruning the tree, through taxation, must ultimately result in smaller annual yields. This example illustrates the concept of capital gains-related disinvestment.

Fears of disinvestment suggest that *nominal* gains should not be taxed. If capital gains must be taxed, then taxes should be applied only to *real* gains. This position requires some mechanism for the separation of nominal gains into their real and inflationary components. How is this to be accomplished?

Inflation-indexing is present in other areas of contemporary tax law (e.g. personal exemptions and standard deductions covered in Chapter 6). However, administrative difficulties may arise in the *equitable* separation of *nominal* gains into their *real* and *inflationary* components. Should the presumed rate of inflation for a specific calendar or tax year be the same for all categories of assets? And should it be permissible for this calculation to result in the generation of an inflation-adjusted tax deductible loss? This represents an administratively expedient and equitable solution, even though resulting calculations could easily lead to reduced tax revenues.

(2) Should Capital Gains Be Taxed Annually, Upon Realization, or at Some Other Time?

How should capital gains be taxed? Capital gains are taxed only upon realization through sale or other forms of disposition, contributing to a modification of taxpayer behavior referred to as "lock-in" effects. Unlike taxable salaries, the taxpayer with the ability to defer consumption arising from capital gains generating investments also possesses the ability to control the date of realization (and taxation).

The alternative, a form of accrual taxation, would be administratively impractical. It would involve the development of an equitable methodology for the cost effective annual valuation of affected assets. It would also involve the inequity of requiring cash outlays to pay related tax liabilities without the benefit of gain-related cash inflows (Pechman, 1971, 96).

(3) Should Capital Gains Be Taxed at Progressive Tax Rates?

Should fully progressive tax rates apply to capital gains? If so, an argument for income-averaging (e.g. 1964–1986 as discussed in Appendix 3C) presents itself. Alternatively, progressive tax rates are a function of the presumed equity of our ability to pay individual federal income taxation.

Current tax law (1999) provides for individual federal income tax rates at 15, 28, 31, 36, and 39.6%. Capital gains are more favorably taxed than ordinary income. The maximum capital gains rate is 20 (reduced to 10 if otherwise taxed at a 15% rate), 25 (for IRC section 1250 property), or 28 (for collectibles, net STCL, and LTCL carryover) percent, depending on the character of the gain (e.g. collectibles taxed at a maximum rate of 28%). Clearly, contemporary tax law provides for the preferential treatment of capital gains.

(4) Should Capital Gains Escape Taxation Through Gift or Bequest At Death?

Should the tax-free transfer of unrealized capital gains occur upon death? A wealthy taxpayer can preserve familial wealth by transferring appreciated assets *after* the taxpayer's death. This transfer results in a "step-up" of the tax basis of the assets to fair market value. This step-up in tax basis may, therefore, result in an effective tax rate of zero. David (1968, 225) noted, with respect to conference attendees:

> A consensus was reached at the conference that the zero rate of taxation at death due to step-up of basis is a major cause of lock-in . . . all agreed that it increases as an investor becomes very old.

Some of the alternatives proposed included (1) *presumptive realization* of taxable income at the time of death, providing tax deferral as the only incentive to hold assets until death,[92] or (2) *carry-over of basis* to the heirs, already applied under present law in the case of *inter vivos* gifts.

(5) Should Capital Losses Be Permitted as Partial or Full Offsets to Ordinary Income?

An unlimited deduction of losses would provide the potential for a complete offset of ordinary income from other sources. In effect, the U.S. Government would be providing economic incentives for increased risk-taking. This would also provide high-income taxpayers with the ability to realize losses in such a manner as to average income over a period of several years. Finally, unlimited capital loss deductions could result in a significant reduction in tax revenues.

Under current law, a ceiling remains in effect and net losses are limited to $3,000 per household, per tax year.

(6) How Does Capital Gains Taxation Affect Risk and Liquidity in Capital Markets?

Opponents of capital gains taxation suggest that low or zero tax rates on capital gains provides the necessary incentives for the *risk* accepted by investors. Capital gains taxes reduce the *liquidity* and free-flow of capital to its "highest and best" use, by imposing a large transaction cost that promotes a "lock-in" effect for the existing investment vehicle. This position has found broad support in a long history of extensive research, conducted under a variety of tax regimes, and published in the academic accounting, finance, and economics literature streams (Cataldo & Savage, 2000).

In mid-April (2000) the stock markets experienced a "correction" or "crash". The major factor was the release of "bad news" in the form of the release of information suggesting higher inflation. However, other factors were involved, including *tax (estimated tax) payment effects*, related to 1999 calendar capital gains realizations. This topic is discussed in Appendix 10N.

THE LEGISLATIVE INTENT OF TAX-FAVORED LONG-TERM CAPITAL GAINS

There are several reasons for the preferential treatment of capital gains (David, 1968, 37):

(1) Securing more equitable tax treatment for investment gains accruing over long periods and which would be assessed in a single tax year under progressive tax rates,
(2) Reducing the inequitable taxation of increments to capital that arise from illusory revaluations (e.g. inflation),
(3) Minimizing interference with the operations of asset markets,

(4) To stimulate investment and spur economic growth, and
(5) To limit the progressive nature of progressive taxation (depends on the
     pattern or trend occurring during the tax period – e.g. declining marginal
     rates).

PUBLISHED EVIDENCE FROM THE CONGRESSIONAL BUDGET OFFICE – 1988

A Congressional Budget Office (CBO) study of data for the years 1954–1985 concluded that *higher tax rates lower capital gains realizations* (CBO, 1988, xi). The sole purpose of this report was to address the issue of estimating individual federal income tax revenues. *Cross-sectional* and *time-series* studies of taxpayer changes in capital gains realizations as responses to changes in tax rates on those realizations were investigated.

Cross-section studies compare taxpayer behavior (or the behavior of taxpayer groups) for a single year (or over several years). The CBO study examined the significance of varying marginal Federal income tax rates on differences in capital gains realizations (relative to GNP), while controlling for dividends, total income, age, and family status.

Time-series studies compare taxpayer behavior over time. The CBO study examined the significance of different marginal Federal income tax rates on total capital gains realizations (relative to GNP), while controlling for real income, wealth (as a proxy for the stock of accrued gains), and price levels.

In a more recent study, Burman and Ricoy (1997, 427) found: (1) that high-income taxpayers realized the overwhelming majority of capital gains, (2) over a ten year period (1979–1988), nearly one-third of all taxpayers reported a capital gain on their tax returns, (3) sales of corporate stock accounted for more gains than sales of any other asset, and (4) after adjusting for inflation, most capital gains disappear. Their study included panel data for the 1979–1988 tax years and the 1993 Sale of Capital Assets (SOCA) data base.

The contemporary LTCG tax ceilings of 20% (10% if otherwise taxed at a 15% rate) on capital gains addresses Burman and Ricoy's concerns regarding prior law, which provided for a 28% ceiling for all taxpayers. People in the 15 and 28% income tax brackets did not benefit from the top tax rate of 28% (Burman & Ricoy, 1997, 449).

The Burman and Ricoy study provided insights into the behavior of taxpayers, but this behavior may have changed. They examined a period pre-dating a raging "bull" stock market, Internet and electronic brokerage account growth, the popularity of day-trading, and increased volatility in these "most efficient" of U.S. capital markets. The importance of this topic to both researchers and policy-makers is addressed in Appendix 10N.

HOLDING PERIODS

There would be no need for short-term or long-term holding period distinctions in the absence of tax rate differentials between capital gains and ordinary income. The continued maintenance of holding period distinctions between short-term and long-term capital gains provides an additional incentive for realization deferral. Taxpayer's anticipating lower future marginal FIT rates are provided with a tax incentive to postpone capital gains realization. The postponement of capital gains realization provides the investor with interest-free use of the deferred tax.

The holding period has played an important role in the taxation of capital gains since passage of the Revenue Act of 1921. At that time, short-term and long-term capital transactions were distinguished by a holding period of two years. Appendix Table 9-5 provides a summary of short-term and long-term holding periods for capital gains and losses.

In 1934, the length of the holding period affected the percentage of gain included in income based on the five (holding period-based) classes of assets, as follows:

- Less than or equal to 1 year at 100%;
- More than 1 year, but not more than 2 years at 80%;
- More than 2 years, but not more than 5 years at 60%,
- More than 5 years, but not more than 10 years at 40%; and
- More than 10 years at 30%.

Beginning with the 1938 tax year, three (holding period-based) classes of assets were established, as follows:

- Less than 1 year at 100%;
- More than 18 months, but not more than 2 years at $66\frac{2}{3}$%; and
- More than 2 years at 50%.

In 1942, a six month holding period breakpoint was established. The six month holding period distinction was the same as that applied to company officers in the definition of a "speculative turn" under the Securities and Exchange Act (Wells, 1949).

Beginning with the 1977 tax year, a nine month holding period was required for preferential LTCG treatment. This period was extended to one year, beginning with the 1978 tax year. It remained at this one year breakpoint through June 21, 1984.

From June 22, 1984–1987, the long-term holding period was six months. It was restored to one year, where it remains under existing tax law, beginning with the 1988 tax year.

CAPITAL GAINS AND THE PERSONAL RESIDENCE

Auten and Reschovsky (1997, 223–224) provide a summary of capital gains taxation relating to the taxpayer's personal residence.

Prior to 1951, capital gains from the sale of a residence were provided the same tax treatment as other capital gains. Capital losses from the sale of a taxpayer's residence was not deductible. For tax years beginning with 1951, taxpayers were allowed to postpone capital gains on the sale of their principal residence if the replacement was purchased and occupied within one year before or after the sale (18 months if newly constructed).

In 1964, in an effort to provide tax relief to elderly taxpayers, a once-in-a-lifetime exclusion of up to $20,000 of the gain from sale of a personal residence was provided. In 1976, this exclusion amount was increased to $35,000. The *Revenue Act of 1978* further increased this exclusion to $100,000 and reduced the age of eligibility from 65 to 55. ERTA81 again increased the exclusion, to $125,000. The *Taxpayer Relief Act of 1997* increased this exclusion to $500,000 ($250,000 for single taxpayers and other non-joint returns), and eliminated the "marriage tax penalty" component (see Cataldo, 1996 and Chapter 8).

SUMMARY

In the U.S., capital gains are taxed. They do, however, receive preferential individual federal income tax treatment. This preferential treatment began with the 1922 tax year, and may have been more important in the past, when progressive Federal income tax rates were much higher.

The taxation of *nominal* amounts of capital gains and not fully allowing for the deduction of capital losses may be inequitable. Adjusting capital gains for inflation may lead to the elimination of gains and/or the generation of (deductible?) tax losses. However, failure to tax capital gains, presumably a tax based on the taxpayer's *ability to pay*, may provide the *appearance* of inequity. Our voluntary system of self-reporting might suffer from greater non-compliance if the capital gains tax were completely eliminated.

The realizations of capital gains are under the control of the taxpayer. Tax deferral suggests that tax savings might be maximized when realizations are deferred indefinitely. However, these "lock-in" effects may be less apparent for taxpayers engaging in the "smoothing" of income. Lock-in effects retard economic growth, inhibiting the flow of economic resources (and economic growth) to their "highest and best use".

Contemporary tax law provides for preferential treatment for all taxpayers (e.g. 10% for those at the 15% tax bracket and 20% for taxpayers at the 28% or higher tax rates). Flat tax proposals (see Chapter 4) would eliminate the preferential treatment of capital gains.

Presently, a tax-free "step-up" in basis is, to a limited degree, available for the U.S. estate tax. The complete elimination of capital gains taxation would extend this feature of estate and gift taxation, and result in a zero tax rate of taxation and increased "lock-in" effects. Furthermore, the elimination of U.S. estate and gift taxation (discussed in detail in Chapter 13), regardless of the relative insignificance of the tax revenues generated by estate and gift taxes, might lead to an increase in the *perception* of U.S. tax law inequities.

The elimination or modification of capital gains taxation was not an issue addressed during the 2000 Presidential campaign. Therefore, we expect that the preferential treatment of long-term capital gains with remain a feature of U.S. individual income tax law for the next decade. For the same reasons discussed and applicable to the AMT (see Chapter 9), we do not believe that the complete elimination of the taxation of capital gains will become politically feasible in the next decade. However, flat tax or national sales tax proposals, if established as replacements for the contemporary U.S. income tax, would eliminate these features. Furthermore, an extended period of U.S. budget surpluses will increase the likelihood of capital gains tax elimination. Finally, if the contemporary surge in individual taxpayer participation in capital stock ownership (brought on by the bull markets) continues, the political palatability for some exclusion (or even the complete elimination) of capital gains taxation may become possible.

PART 4

GENERATIONAL TAXATION

CHAPTER 11
SOCIAL SECURITY AND THE
SELF-EMPLOYMENT TAX

... the industry of this generation should pay the bills of this generation (Pollack, 1996, 54, quoting Woodrow Wilson).

"(G)enerational accounting" reveals how much money the government ... is slated to take from each contemporary generation as well as those generations not yet born (Kotlikoff, 1992, ix).

... in 1996 ... the Social Security Advisory Commission ... noted that the current surplus ... would run out in 2012 ... while Americans currently carry an average tax burden of something like 30%, that figure will have to move to 50% if future generations are to sustain the baby boomers in retirement (Shlaes, 1999, 65–66, citing Kotlikoff).

Under current projections of the Board of Trustees of Social Security, the Social Security Trust Fund will be depleted by 2037 (Lyon & Stell, 2000, 473).

Social Security has already accumulated an implicit debt that cannot be fully revoked ... substituting mandatory IRAs for traditional Social Security benefits and taxes would not relieve Social Security's implicit debt burden ... the free lunch argument is rejected ... (Mariger, 1999, 784).

INTRODUCTION

The 1980s saw the broad application of the phrase "revenue neutral". Basically, a tax law or revenue or expenditure policy change was revenue neutral if the net result of the change did not result in any "net" change in the U.S. budget surplus or deficit. A tax increase of $20 billion was matched with an expenditure increase of $20 billion and was, therefore, revenue neutral.

Alternatively, a proposed change that did not affect either revenues or expenditures, but resulted in a redirection of federal economic resources from one group to another (e.g. from younger taxpayers to social security beneficiaries) was also considered revenue neutral. Kotlikoff recommended that a shift in focus take the form of viewing such changes on a "generational" dimension.

In the 2000 Presidential campaign, George W. Bush proposed that a portion of Social Security payments for younger taxpayers be invested in equities. This plan is one of several privatization plans that are in their infancy. Though we believe that a plan of partial social insurance privatization may very well become available in the next few decades, we expect that such plans will, in large part, depend upon a stabilization of the equity markets. Therefore, we do not address these privatization plans, which are beyond the scope of this chapter.[93]

THE SOCIAL SECURITY CRISIS OF THE EARLY 1980s

The U.S. National Commission on Social Security Reform of 1981, ERTA81, and TRA86 resulted in simultaneous: (1) increases in payroll and self-employment tax rates, and (2) decreases in marginal Federal income tax rates. These changes in the relative importance of Social Security and Federal income tax rates led to reduced overall progressivity, but increased taxes for the self-employed taxpayer (Calegari, 1993; Pechman, 1985; CBO, 1987; Ricketts, 1991).

Presently, employers and employees each contribute 7.65% (15.3% in total) to the Social Security (6.2%) and Medicare (1.45%) tax systems. The U.S. tax system refers to this as the Federal Insurance Contributions Act (FICA) tax. Self-employed taxpayers contribute amounts equivalent to both employer and employee portions, or the entire 15.3%. The U.S. tax system refers to this as the Self-Employed Insurance Contributions Act (SECA) tax.

Prior to the 1990 tax year, self-employed taxpayers received a tax subsidy, which meant that they were not required to pay an amount equal to combined employer and employee components. This subsidy was gradually eliminated by the U.S. National Commission on Social Security Reform of 1981, over the 1982–1989 tax years.

For example, the employer and employee each contributed 6.65%, for a total of 13.3% for the 1981 tax year. The self-employed taxpayer contributed only 9.3% for the same year, receiving a 4% subsidy. This subsidy decreased to 2% for the 1986–1989 tax years, and was completely eliminated for the 1990 and all subsequent tax years.

This bipartisan Commission was formed to develop a solution to the (then) anticipated insolvency of the U.S. Social Security and Medicare tax systems, which are not funded from general FIT revenues. Presently, the Social Security and Medicare tax systems are producing surpluses, largely due to the elimination of the subsidies previously enjoyed by the self-employed taxpayer.

THE IMPACT OF RISING SECA TAXES ON SMALL BUSINESS FORMATION AND PROFITABILITY

This chapter contains an exploratory study which uses published data made available by the Statistics of Income (SOI) branch of the Internal Revenue Service (IRS) for the years 1951–1992 to investigate longitudinal relations between the profitability of sole proprietorships and SECA tax rates, which rose over this period. The 1989 IRS SOI Public Use File from 1992 is used to test this and to further explore cross-sectional relations between the profitability of sole proprietorships and marginal FIT rates.

Results suggest that the percentage of individual taxpayers engaging in and reporting the results of their self-employment endeavors has not been significantly affected by rising SECA tax rates. However, the percentage of sole proprietors reporting taxable profits has declined from 1951–1992. This decline bears a strong and statistically significant relation to rising SECA tax rates.

The independent and control variables used for the longitudinal model maintain their correlational relations for a cross-sectional analysis using data from 1989. There is no apparent decline in sole proprietorship profitability as taxpayers' marginal FIT rates increase in the later analysis.

These findings are intuitively appealing, supported by the literature, and consistent with rational tax minimization strategies and/or a broad longitudinal awareness of the net results of: (1) increasing SECA tax rates, and (2) decreasing marginal FIT rates evolving from ERTA81 and TRA86.

HISTORICAL BACKGROUND

The U.S. Social Security Act was signed into law on August 14, 1935. It provided separate programs for an old-age retirement benefits system and an unemployment insurance system. A response to the Great Depression, Social Security was established to compensate for the destruction of the savings of the elderly and the reductions in employment opportunities characteristic of the period. Taxes were first collected during the 1937 tax year. Monthly benefits first became available in 1942 (SSA, 1987).

The first FICA tax applied to employers and employees and provided for a 2% combined employer/employee contribution, beginning with the 1937 tax year. Sole proprietors were not affected at this time. The SECA tax applied only to sole proprietors. It was not imposed until the 1951 tax year. SECA tax benefits were not extended to sole proprietors until the early to mid-1950s.

120 ANTHONY J. CATALDO II AND ARLINE A. SAVAGE

THE SECA TAX SUBSIDY AND PHASE-OUT (1951–1989)

The early years of the SECA tax (1951–1953) provided for an initial rate of 2.25% of self-employment profits of four hundred dollars or more. The combined employer and employee contribution for FICA was 3% for the same period. Therefore, a "subsidy" of 0.75% (3% less 2.25%) was given to the sole proprietor.

For the 1951–1989 tax years, SECA taxes were imposed at subsidized rates. This subsidy reached a maximum of 4.16% for the 1979 and 1980 tax years. It was phased out and completely eliminated beginning with the 1990 tax year.

Inflation, slow growth in wages, and unemployment during the late 1970s and early 1980s, combined with declining birth rates and increasing life expectancy, led to the formation of the National Commission on Social Security Reform in 1981. This bipartisan Commission made recommendations to increase Social Security program financing. These recommendations, leading to 1983 Amendments to U.S. tax law, included an acceleration of tax rate increases previously scheduled, as well as permanent increases (including the subsidy phase-out) in SECA tax rates.

DEDUCTIBILITY OF SECA TAXES (1990–)

Beginning with the 1990 tax year, the introduction of the adjustment to adjusted gross income (AGI) feature for 50% of the SECA tax reduced the after-tax cost from its 15.3% pre-tax level. This feature made the SECA tax deductible for sole proprietors. The FICA tax is not (and never was) deductible for employees.

For the post-1989 tax years, the after-tax cost of SECA taxes can be computed using equation (11-1), as follows:

$$[0.153 \times (1.0 - 0.0765) \times (1.0 - 0.5T)] \tag{11-1}$$

where 15.3% is the gross SECA tax rate prior to AGI adjustment for a reduction of self-employment income for SECA tax computation purposes (e.g. 100% less 7.65% equals 92.35%). This deduction of 50% (e.g. 15.3% multiplied by 50% equals 7.65%) of the SECA tax results in a marginal FIT rate (T) dependent reduction in total taxes.[94]

HISTORICAL TABLES AND FIGURES – APPENDIX

The entire history of FICA and SECA tax rates is contained the Appendix Tables. Appendix Table 11-1 contains historical pre-tax FICA (1937–) and

SECA (1951–) tax rates. Post-tax SECA tax rates, using equation (11–1), are also provided in Appendix Table 11-1. Appendix Table 11-2 summarizes maximum FICA (1937–) or SECA (1951–), Medicare, and total contribution amounts. Appendix Tables 11-3 (1951–1989), 11-4 (1951–1989), and 11-5 (1966–) more fully illustrate SECA tax-related measures. Some of these measures are discussed and used in the regression models developed later in this chapter.

Related figures are also provided. Appendix Fig. 11-1 provides a graphical comparison of FICA and SECA tax rates (1951–1989). Appendix Fig. 11-2 focuses on the SECA tax subsidy component for the same period. Appendix Fig. 11-3 expands on the SECA tax component contained in Appendix Fig. 11-2, graphically illustrating this subsidy in both nominal and CPI-adjusted forms. Appendix Fig. 11-4 provides a comparison of weighted average AGIs for all taxpayers (All) and taxpayers with SECA taxes (SE) only. (Note that the weighted average AGI for self-employed taxpayers has increasingly lagged behind the AGI for all taxpayers.) Finally, Appendix Fig. 11-5 illustrates the growth in the FICA (and SECA) wage base (1937–).

CONTEMPORARY RESEARCH

Calegari (1993) concluded that most self-employed taxpayers were worse off and bore a larger tax burden in 1990 than either 1980, 1970, or 1960. Other studies of periods inclusive of the ERTA81 and TRA86 have found reduced overall tax progressivity during the 1966–1988 period (Pechman, 1985; CBO, 1987; Ricketts, 1991). Although these Acts may have contributed to general increases in horizontal equity[95] (Ricketts, 1991), the self-employed represent a class of taxpayers (relatively) adversely affected by this growth in Social Security, Medicare, and related SECA taxes.

EXPERIMENTAL RESEARCH

Recent experimental evidence suggests that " . . . most taxpayers are unable to provide an accurate estimate of their (marginal FIT rate) . . ." (Rupert & Fischer, 1995, 38). These results were based on a survey instrument distinguishing between Federal and state income taxes. Rupert and Fischer concluded that respondents' perceptions of their marginal FIT rates were (generally) overstated. However, their instrument did not specifically address SECA taxes.

Their results indicate that 18.5% of their survey respondents filed a Schedule C or Schedule F with their individual FIT return. The Schedule C is filed by

U.S. taxpayers with non-farm self-employment profits or losses. The Schedule F is filed by U.S. taxpayers with self-employment profits or losses from farming or ranching endeavors. These Schedules are attached to the U.S. taxpayer's return (Form 1040).

Eighty percent of the sole proprietor survey respondents attached a Schedule SE to their FIT return. The Schedule SE is only attached to the U.S. FIT return when the SECA tax applies. The SECA tax applies only when there are taxable profits from self-employment. Farming or non-farming sole proprietorships with losses are required to file the Schedule F or C, respectively, but are not required to file the Form SE (or pay SECA taxes) if the self-employment activity results in a loss.

The majority of their respondents (58.3%) had actual marginal FIT rates of 15%. This rate is slightly below that of the current pre-tax SECA rate of 15.3%. Only 23.1% of the respondents estimated their own marginal FIT rates to range between 0 and 15%. Therefore, these taxpayers overstated their marginal FIT rates.

In commenting on Rupert and Fischer, Collins (1995) suggested that state income tax rates might have been included in the respondents' "perceived" marginal FIT rates.[96] The instrument did not provide for an adequate distinction between FIT and state income tax rates. Perhaps some of these taxpayers also failed to distinguish between declining FIT rates and rising FICA/SECA tax rates. In the U.S., FIT and SECA taxes are: (1) calculated on the same FIT return (Form 1040) at year-end, and (2) are paid for in a combined form with a single transmittal voucher and check.

RESEARCH QUESTIONS, METHODOLOGY, AND RESULTS

Have declining individual FIT rates been offset by the rise in SECA tax rates as Calegari (1993) concluded? A preliminary analysis, based only on maximum FIT rates, is consistent with his position. Correlating maximum FIT rates to SECA tax rates for 1951–1995 produces a Pearson product-moment correlation coefficient of -0.96 ($p < 0.0001$).

Figure 11-1 provides visual evidence of the inverse relations between rising SECA tax rates (from Appendix Table 11-1) and decreases in the percentage of sole proprietors' reporting taxable earnings from self-employment (from Appendix Table 11-3). Note that the percentage of individual federal income tax returns containing a Schedule C or F (from Appendix Table 11-3) have remained relatively stable over this period. A Pearson product-moment correlation between these two measures is 0.72 ($p < 0.0001$).

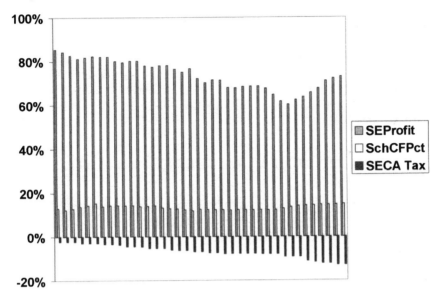

Fig. 11-1. Percent of SE Taxpayers with Profits, Percent of Individual Federal Income Tax Returns with Schedules C and F, and the SECA Tax Rate (negative) (1951–1989).

Unlike employees, it is possible for self-employed taxpayers to manipulate, through legal tax planning strategies, the amount of SECA taxes to which they are subject. For example, additional business deductions reduce FIT, state income tax, and SECA tax for the self-employed sole proprietor, but do not result in a reduction of FICA taxes (the employees' counterpart to the sole proprietor's SECA tax) for the employee. Therefore, the economic value of the same deduction or dollar reduction in taxable earnings from self-employment (U.S. FIT Form 1040 Schedules C or F earnings for the non-farm and farm sole proprietor, respectively) differs between employees and sole proprietors.[97]

Research Questions: Given the increasing relative importance of SECA taxes, does evidence support the position that small business formation or profitability is inhibited or alternatively that self-employed taxpayers are engaging in tax minimization strategies?

DEPENDENT VARIABLES INVESTIGATED

Two separate measures of self-employment are investigated. The first is the percentage of taxpayers filing Schedules C (non-farm) and F (farm) with their annual FIT return (PCTCF). The second is the percentage of taxpayers reporting profits on their Schedules C and F (PROFPCT). The three independent (control) variables included in the regression models used to examine relations between these two dependent variables and rising SECA tax rates (SETAXRT) are described below. (The independent variable measures used for the below regression equations are contained in Appendix Tables 11-1–11-4.)

INDEPENDENT VARIABLES INVESTIGATED

Sole proprietors' taxable earnings/profits from self-employment activities are subject to SECA, FIT, and (if applicable) state income tax. Unlike wages, profits (or losses) from self-employment are subject to manipulation and, therefore, tax-minimization. This ability to reduce or eliminate SECA taxes represents an option not available to employees for their FICA tax counterpart. (The dependent variable measures used for the below regression equations are contained in Appendix Tables 11-1–11-4.)

The SECA Tax Rate

Increases in SECA tax rates (which generally occurred over the research period) are expected to correlate with a decline in the number of taxpayers willing to engage in profitable self-employment endeavors. Similarly, SECA tax rate increases produce increased economic incentives for affected taxpayers to engage in legal tax-minimization strategies, directed toward the reduction (or elimination) of profits from self-employment.

U.S. Internal Revenue Code (IRC) Section 179 Expense Election

The IRC Section 179 expense election, first introduced for the 1958 tax year, provides a vehicle for SECA tax reduction. It was designed specifically for the purpose of promoting small business formation and expansion through accelerated depreciation. Therefore, all else being equal, increases in SECA tax rates provide increased incentives for IRC Section 179 qualifying capital investments (Scholes & Wolfson, 1992), where the election of IRC Section 179 is nearly always preferable (Auster, 1988; Holtz-Eakin, 1995).

The Consumer Price Index

Empirical evidence supports an inverse relation between corporate capital investment and debt financing (Dhaliwal, Trezevant & Wang, 1992). Furthermore, empirical evidence suggests that capital market constraints (that in, higher relative nominal interest rates) reduce the formation of new businesses and lower the survival rates among less established firms (Holtz-Eakin, 1995; Holtz-Eakin, Joulfaian & Rosen, 1994a, b; Evans & Leighton, 1989; Evans & Jovanovic, 1989). This study uses the Consumer Price Index (CPI) as a consumer/sole proprietor-based proxy for the inflation rate component of nominal interest rates. All else being equal, higher measures of CPI are expected to inhibit small business formation, expansion, and profitability.

The Earned Income Credit (EIC)

This feature of U.S. tax law represents a tax return-based form of welfare for the "working poor". The manipulation of sole proprietor taxable earned income provides a vehicle for EIC maximization (O'Neil & Nelsestuen, 1994). This may be achieved through use of the IRC Section 179 election as discussed above (Cataldo, 1995). Some may even declare self-employment when none has taken place in order to receive the EIC (Steuerle, 1993). Because both phase-in and phase-out ranges exist for the EIC, no prediction will be made regarding the desirability of reporting self-employment profits and losses.

Cataldo (1995) provides a very detailed summary, complete with graphic representations, of the contemporary EIC (1975–). The EIC is similar to a negative income tax, guaranteeing some minimal level of income to all U.S. taxpayers, but is more appropriately characterized as "workfare" rather than "welfare". A working taxpayer with earned income may qualify for the EIC. A non-working taxpayer without earned income cannot qualify for the EIC. Therefore, the EIC may provide an economic incentive for low income taxpayers to work. It may also provide an economic incentive for sophisticated U.S. taxpayers to legally manipulate their earned income.

Earned income is a "net" measure and includes profits and losses from self-employment, as well as wages. The EIC has a: (1) phase-in, (2) flat, and (3) phase-out range. These ranges are linear. They are inflation-indexed and change from tax year to tax year.

The EIC is a refundable tax credit. It is maximized in its "flat range". Therefore, U.S. taxpayers may choose to work more or less to position their earned income within this flat range. There are, of course, other considerations, as this issue is a complex one. For example, how does one distinguish between sophisticated and unsophisticated U.S. taxpayers, capable of manipulating their earned income into the EIC-maximizing flat range? For this reason, the development of an all-inclusive model to deal with the EIC control variable was not attempted and is beyond the scope of this study.

A LONGITUDINAL ANALYSIS OF SOLE PROPRIETORS (1951–1992)

How have rising SECA tax rates affected sole proprietorship formation and profitability? What relations, if any, exist among changes in the maximum amounts available for the IRC Section 179 expense election, the CPI, increasing EIC tax rate percentages, and the formation and profitability of sole proprietorships?

Equation (11-2a) was used to investigate the predicted relations between these variables and the percentage of total individual taxpayers filing Schedules C and/or F with their Forms 1040 for the 1951 through 1992 tax years, as follows:

$$PCTCF_i = \beta_0 + \beta_1 SETAXRT_i + \beta_2 IRC179_i + \beta_3 CPI_i + \beta_4 EICPCT_i + \varepsilon_i \quad (11\text{-}2a)$$
$$(-) \qquad\qquad (+) \qquad (-) \qquad\quad (?)$$

Equation (11-2b) was used to examine similar predictions and the same independent variables, but with respect to their relationships to reported profits, as follows:

$$PROFPCT_i = \beta_0 + \beta_1 SETAXRT_i + \beta_2 IRC179_i + \beta_3 CPI_i + \beta_4 EICPCT_i + \varepsilon_i \quad (11\text{-}2b)$$
$$(-) \qquad\qquad (+) \qquad (-) \qquad\quad (?)$$

where

   PCTCF = the total number of Schedules C/F filed as a percentage of all individual Federal income tax Forms 1040 filed,

SETAXRT = the (net) SECA tax rate before post-1990 after-tax adjustments,

   IRC179 = the maximum Section 179 expense election available per return,[98]

     CPI = the annual rate of inflation as measured by the CPI,

  EICPCT = the maximum EIC percentage available for the tax year, and

PROFPCT = the percentage of Schedules C/F reporting profits for tax year i, ranging from 1951 through 1992.

Table 11-1 contains the results of regression Eqs (11-2a) and (11-2b.) Data are available from the author on request.

Table 11-1. Results of Regression Models for Equations (11-2a) and (11-2b) (t-values in parentheses).

DV: Equations ⇒	PCTCF:Equation (2a)		PROFPCT: Equation (2b)	
IVs	Predicted Sign	Regression Result	Predicted Sign	Regression Result
Intercept	?	0.14**** (20.07)	?	0.88**** (26.65)
SETAXRT Self-Employment Tax Rate	–	–0.08 (–1.00)	–	–2.60**** (–6.49)
IRC179 IRC Section 179 Expense Election	+	1.25x10^6 (1.50)	+	1.60x10^{-5}*** (3.99)
CPI Consumer Price Index	–	–0.20**** (–4.91)	–	–0.50* (–2.56)
EICPCT Earned Income Credit Percentage	?	0.08** (2.91)	?	0.12 (0.92)
N		42		42
Adjusted R^2		0.59		0.76
F for Regression		15.72****		33.49****

**** $p < 0.0001$
*** $p < 0.001$
** $p < 0.01$
* $p < 0.1$

As predicted, for Eq. (11-2a) only the SETAXRT (n.s.) and CPI (significant at the 0.00001 level) independent variables were inversely related to the percentage of taxpayers filing Schedules C and/or F (PCTCF). The relationship between IRC179 and PCTCF was in the predicted positive direction. The EICPCT was also positively related to PCTCF, and was statistically significant at the 0.01 level.

Equation (11-2b) examines the relationships between the percentage of Schedules C and/or F reporting profits (PROFPCT) and the three independent variables. Hypothesized coefficient signs were as predicted. However, the SETAX variable is now significant, suggesting that rising SECA taxes have resulted in a decline in the percentage of sole proprietors reporting profits. Similarly, IRC179 is now significant at the 0.001 level. The relationships of both SETAXRT and IRC179 to PROFPCT are intuitively appealing and supported by prior corporate research findings (Scholes & Wolfson, 1992).

The combined results from Fig. 11-1 and Eqs (11-2a) and (11-2b) suggest that rising SECA tax rates (SETAXRT) have not resulted in any reduction in the percentage of individual taxpayers filing Schedules C/F with their Form 1040, but have resulted in a reduction in the percentage of Schedules C/F reporting profits (PROFPCT).

EVIDENCE FROM 1989 CROSS-SECTIONAL IRS SOI DATA

The 1989 tax year represented the final year of the SECA tax subsidy phase-out. For the 1989 tax year, using a regression model comparable to those employed by Eqs (11-2a) and (11-2b), all taxpayer records with SECA taxes and positive earnings were extracted from the 1989 IRS SOI. Table 11-2 contains a summary of the population and sample data used for this portion of the study.

Equation (11-2b) used the percentage of tax returns reporting positive net earnings on Schedules C and F for the 1951 through 1992 tax years. Similarly, the regression model developed for the single 1989 tax year focuses on tax returns with positive net earnings (and, therefore, resulting in some positive amount of SECA tax). As Table 11-2 indicates, 15,309 of the returns in the 1989 IRS SOI possessed these characteristics and were selected for testing.

Unlike the longitudinal designs used for Eqs (11-2a) and (11-2b), the 1989 IRS SOI data, consisting of FIT return information from a single tax year,

Table 11-2.   Summary of Samples Drawn from 1989 IRS SOI Data.

Total Federal income tax returns in 1989 Individual IRS SOI	96,588
Less: Federal income tax returns without SECA tax	77,256
Less: Federal income tax returns without positive net earnings from combined Schedules C/F	4,023
Equals: Federal income tax returns with SECA tax and positiveEqualsFederal income tax returns with SECA tax and positive net earnings from combined Schedules C/F	15,309
Less: Federal income tax returns with AGIs above (below) $100,000 ($1)	4,495
Equals: Federal income tax returns with AGIs between $1 and $100,000	10,814

does not provide for differing SECA tax rates. However, different marginal FIT rates are provided. Because net earnings from self-employment contributes to increases in taxable income and marginal FIT (MFIT) rates, it is hypothesized that higher self-employment net earnings will have a positive relationship with marginal FIT rates.

Similarly, the 1989 IRS SOI provides measures of depreciation (DEPR) expense[99], interest (INT) expense,[100] and taxpayer earned income credits (EIC). These measures were used as proxies for IRC179, CPI, and EICPCT independent variables respectively, contained in Eqs (11-2a) and (11-2b). The hypothesized directions for these independent variables remain unchanged from Eqs (11-2a) and (11-2b).

Equation (11-2c) provides for this conversion and predictions for directions of the four independent variables, as follows:

$$SCHCF_i = \beta_0 + \beta_1 MFIT_i + \beta_2 DEPR_i + \beta_3 INT_i + \beta_4 EIC_i + \varepsilon_i \quad (11\text{-}2c)$$
$$(+) \qquad (+) \qquad (-) \quad (+)$$

where

SCHCF = the (net) earnings reported on tax returns with both SECA taxes and positive (net) earnings for combined Schedules C/F,
MFIT = the taxpayer's marginal Federal income tax (FIT) rate,
DEPR = the depreciation expense reported on the taxpayer's Schedule C,
INT = the interest expense reported on the taxpayer's Schedule C, and
EIC = the EIC reported on the taxpayer's Federal income tax return.

Equation (11-2c) was estimated, separately, for: (1) the full 1989 IRS SOI sample (N = 15,309), and for (2) a subsample (N = 100), consisting only of those tax returns with AGI classes ranging from $1 through $100,000. The latter used the mean measures produced separately for each AGI class from $1,000 to $100,000, representing approximately 70% (10,814; see Table 11-2) of the observations contained in the 1989 IRS SOI.[101] Table 11-3 contains a summary of the results from regression Eq. (11-2c).

Like Eqs (11-2a) and (11-2b), the signs of coefficients for depreciation expense, interest expense, and earned income credit are the same under both variations of Eq. (11-2c). A summary of the signs of coefficients from all three regression equations, including both variations of Eq. (11-2c) is presented in Table 11-4. These results suggest that SECA tax rates have an inverse relationship to self-employment taxable profits.

Table 11-3. Results of Regression Models for Equation (11-2c) (t-values in parentheses).

DV: SCHCF				
IVs	Predicted Sign	All AGIs	$1–$100K	Coefficient Sign
Intercept	?	–111,360 (–0.3)	–10,398 (–1.2)	Not hypothesized.
MFIT Marginal Federal Income Tax Rate	+	10,526**** (36.8)	1,143**** (7.8)	As Predicted for both model variations.
DEPR Depreciation Expense	+	1.4**** (23.1)	4.0**** (5.9)	As Predicted for both model variations.
INT Interest Expense	–	–0.5 (–0.7)	–0.8 (–0.9)	As Predicted for both model variations.
EIC Earned Income Credit	+	110**** (5.1)	12.2 (0.9)	As Predicted for both model variations.
N		15,309	100	
Adjusted R^2		0.13	0.74	
F for Regression		602.4****	72.0****	
Percent of Population		100	90.82	
Percent of Sample		100	70.06	

**** $p < 0.0001$
*** $p < 0.001$

Limitations

This exploratory study examined U.S. sole proprietorships. No attempt was made to conduct comparable analyses of U.S. partnerships, trusts, or corporations. In fact, the earned income credit (EIC) does not apply (directly) to the taxation of U.S. partnerships or trusts, and has no applicability to U.S. corporations. Similarly, the Internal Revenue Code (IRC) Section 179 expense election (IRC179) does not (directly) apply to the taxation of U.S. partnerships or trusts, and applies only to the smallest of U.S. corporations. Finally, the

Table 11-4. Summary of Coefficient Signs for Independent Variables Used in Regression Equations (11-2a), (11-2b), and (11-2c).

IVs	Dependent Variable (DV) – Regression Equations				
	Longitudinal		Cross-Sectional (2c)		
	PCTCF (2a)	PROFPCT (2b)	SCHCF All AGIs	SCHCF $1–$100K	Relation
Not Consistent:					
SETAXRT Self-Employment Tax Rate	–	–			Negative
MFIT Marginal Federal Income Tax Rate			+	+	Positive
Consistent:					
IRC179 IRC Section 179 Expense Election	+	+			Positive
DEPR DepreciationDepreciation Expense			+	+	Positive
CPI Consumer Price Index	–	–			Negative
INT Interest Expense			–	–	Negative
EICPCT Earned Income Credit Percentage	+	+			Positive
EIC Earned Income Credit			+	+	Positive
N	42	42	15,309	100	

SECA tax applies to U.S. partnerships and trusts only indirectly and/or infrequently, and then applies only to a very specific form of U.S. corporation, referred to as the "S corporation." The S corporation is frequently associated with closely held corporations owned and controlled by families and not listed on national or regional stock exchanges. Therefore, the generalizability of these results to alternative U.S. organizational forms may not be appropriate, despite the largely successful Congressional efforts to eliminate the horizontal inequities or the tax advantages and disadvantages associated with their selection.

Concluding Remarks and Suggestions for Further Study

This exploratory study used historical (1951–1992) data, provided by the Internal Revenue Service's Statistics of Income Division – Individuals, to develop a model and explore and test the relationship between two separate measures of sole proprietorship profitability and increases in self-employment tax rates. Cross-sectional data for the 1989 tax year was then used to test comparable independent and control variables for the directional consistency of coefficients.

After controlling for the effects of inflation (nominal interest expense), rising depreciation expense ceilings (total depreciation expense), and rising earned income tax credit rates (dollar amounts), statistical evidence suggests that historical increases in self-employment tax rates are inversely related to: (1) sole proprietor formation/retention, and (2) the percentage of sole proprietors reporting/generating taxable profits. These results are consistent with recent historical shifts in the relative importance of rising (declining) self-employment (Federal income) tax rates.

The past two decades have seen an increase in the relative importance of U.S. payroll (FICA and SECA) tax rates. Over the same period, U.S. individual Federal income tax (FIT) rates have declined. This study has focused on the most affected group of individuals, the sole proprietor or self-employed U.S. taxpayer. In some cases and for the first time in U.S. history, a small business taxpayer's SECA tax rate of 15.3% may exceed his/her marginal Federal income tax rate of (0 or) 15%!

This exploratory study investigated the relevance of rising SECA tax rates to small business/sole proprietorship formation and profitability. SECA tax rate increases were designed to ensure the solvency of the U.S. social insurance system and resulted from recommendations made by the U.S. National Commission on Social Security Reform of 1981. The impact of declining FIT rates was also examined. Decreasing marginal FIT rates resulted from ERTA81 and TRA86.

Longitudinal (1951–1992) analyses were conducted to separately examine the impact of rising SECA tax rates on small business (1) formation and (2) profitability. Rising SECA tax rates were indirectly related to both small business formation and profitability, but only those associated with sole proprietorship profitability were statistically significant (see Table 11-4) with respect to the SECA tax variable of interest (SETAXRT). These results provide evidence that small businesses have used legally available U.S. tax planning strategies to reduce the profitability of their sole proprietorships and, therefore, the impact of rising SECA tax rates.

Cross-sectional (1989) analyses were conducted to increase the rigor of the study and to determine the directional consistency of proxies representing control variables used for the longitudinal analyses. Two variations were developed for the cross-sectional analysis, which focused only on U.S. taxpayers with small business profits (see Table 11-3). Results were consistent with respect to the proxies for the control variables used in the longitudinal analyses (see Table 11-1). As predicted, the marginal FIT (MFIT) rate produced a significant and direct relation to sole proprietorship profitability. U.S. maximum MFIT rates had decreased and additional reductions were scheduled for future tax years.[102]

What are the tax policy or international implications of these results? Nations considering tax policy changes, specifically those affecting sole proprietorships, will find the results of this study of interest. Similarly, small business communities will find these results useful to assist their governments in tax policy formulation.

For example, in contrast to the U.S. system, the Canadian system of taxation does not provide for an additional component of tax on the earnings of Canadian sole proprietorships. The Canadian social insurance system is funded from general tax revenues. What would happen if the Canadian system of taxation were altered to provide for an additional component of taxation on the Canadian small business or sole proprietorship? Would Canadian small business/sole proprietorship formation and profitability decline?

The present study provides results from the examination of U.S. small businesses over the 1951–1992 tax years. A similar study using Canadian data, or that from any other nation where similar tax rate increases were or were not imposed, would provide further insight into the impact of rising U.S. SECA tax rates on U.S. small business formation and profitability. The study contained in this chapter provides the first step toward further investigation of this topic.

SUMMARY

The solvency of the U.S. Social Security system is likely to come under pressures during the next two decades. During the 2000 election year, it became politically popular to suggest that all or some of the Social Security contributions should come under the control of the taxpayer-beneficiary. With the benefit of restrictions that have yet to be fully developed, the objective was to provide the taxpayer with access to the high stock market returns, like those occurring during the mid- to late-1990s. These discussions did not cease with the April 2000 "crash" (see Appendix 10N).

It is quite possible that anticipated Social Security system solvency issues will be resolved as they have been in the past. A U.S. National Commission on Social Security Reform, like the one formed in 1981, will be established to resolve the economic shortfall associated with the changing demographics. Under this scenario, Social Security payroll tax rates and wage bases will be increased to resolve future solvency issues.

Alternatively, a stabilization of the equity markets and continued economic growth would lead us to anticipate growing favor toward at least partial privatization of Social Security contributions. Chapter 12 addresses deferred compensation plans. We view this as a companion topic, since any voluntary or partial privatization of Social Security contributions would be comparable to existing deferred compensation plans.

CHAPTER 12
DEFERRED COMPENSATION

The (tax) advantages given to . . . "qualified plans" are intended to help assure adequate retirement incomes . . . stimulate national saving and economic growth (CBO, 1987, xi).

INTRODUCTION

Deferred compensation plans have enjoyed tremendous growth and popularity in recent decades. For example, when the "active participant" restrictions on U.S. individual retirement account (IRA) contributions were relaxed during the 1982–1986 tax years, contributions soared. The desire for tax deferral was so great, particularly among high-income taxpayers, that tax revenues declined significantly. Further restrictions on contributions for taxpayers already participating in employer plans were re-imposed for the 1987 and subsequent tax years.

During the 2000 Presidential campaign, Bush's (R) proposals included the expansion of education IRAs and the (initial) privatization of some component of Social Security (see Chapter 11) contributions. Vice-President Gore (D) proposed a new tax-sheltered retirement savings account, but with government matching funds for incomes up to $100,000 on joint returns and $50,000 on single returns. Similar privatization proposals were made in Canada's 2000 election for their Prime Minister. Their Reform Alliance Party proposed the diversion of Canadian Pension Plan (CPP) contributions to a registered retirement savings plan (RRSP). The Reform Alliance Party, a conservative party that might be compared with the Republican Party in the U.S., was unsuccessful in this election. Because some form of private retirement plan is popular with both U.S. political parties, we anticipate the near-term passage of some hybrid of these proposals.

This chapter provides a very brief review of the historical and contemporary background information on both U.S. and Canadian deferred compensation plans. With respect to the U.S. IRA, some exploratory statistical evidence is also provided. We conclude the chapter with some additional observations relating to long-term U.S. tax policy on deferred compensation.

HISTORICAL BACKGROUND

Deferred compensation includes both *nonqualified* and *qualified* plans. A very brief history of both follows:

Nonqualified plans are primarily a tax deferral device for executives:

- WWII growth occurred when individual Federal income tax rates were at their highest historical levels;
- TRA69 brought about a decline in individual federal income tax rates (and the popularity of tax deferral) though stock options and other restricted property plans became popular after the establishment of IRC Section 83; and
- TRA86 reduced individual federal income tax rates further, but new vehicles like the Rabbi and Secular trust became popular.

Qualified plans have a longer history and broader applicability:

- During the 19th century, white collar pensions originated;
- During 1924, the IRC established non-discrimination provisions for highly compensated employees;
- During the 1930s and 1940s, the establishment of Social Security led to the establishment of Union defined benefit pension plans;
- During 1942, the IRC revised anti-discrimination provisions;
- During 1962, qualified plans for self-employed individuals were established;
- ERISA74 resulted in the overhaul of IRC Sections 401–419, providing for the establishment of Individual Retirement Accounts (IRAs); and
- TEFRA82 was designed to roll back some of the excesses of ERISA74, increase Federal income tax revenues, and coordinate *qualified plans*; and resulted in the required reduction of most qualified benefits.

CONTEMPORARY SMALL BUSINESS RETIREMENT PLANS

A variety of deferred compensation or retirement plans are provided for under U.S. tax law. Some mention of all of these plans is warranted.

Simplified employee pensions (SEPs) provide a vehicle for employers contributing to their employee's retirement plans. The SEP is frequently used in lieu of profit sharing or money purchase plans with a trust. The employer simply adopts a SEP agreement and makes contributions directly to an IRA established for each eligible employee.

The Savings Incentive Match Plan for Employees (SIMPLE) plan can be established by employers with 100 or fewer employees with earnings above $5,000. Employees can make salary reduction-based contributions to SIMPLE plans. In addition, employers make matching or non-elective contributions.

Keogh (HR 10) plans are typically employed by self-employed taxpayers. Keogh plan rules are more complex that SEP or SIMPLE plans.

Individual retirement accounts (IRAs) represent personalized savings plans for individual taxpayers. In 1997, a provision was made to permit the establishment of a Roth IRA, where contributions are made on an *after-tax* basis and do not provide for immediate tax savings.

THE U.S. INDIVIDUAL RETIREMENT ACCOUNT

... the likelihood of IRA participation increases as the balance due with the taxpayer's return increases ... consistent with the presence of ***framing bias*** (emphasis added) in taxpayers' decision-making ... preparers appear to reduce framing bias in wealthy persistent savers ... (Frischmann, Gupta & Weber, 1996, 2).

The U.S. individual retirement account (IRA) first became available for the 1975 tax year. The Canadian counterpart, the registered retirement savings plan (RRSP), preceded the U.S. IRA, beginning with the 1958 tax year. Summaries of selected descriptive measures for both the U.S. IRA (1975–1996) and the Canadian RRSP (1968–1995) are contained in Appendix Tables 12-1 and 12-2. (Summaries of descriptive measures of U.S. Keogh or SEP plan contributions are provided in Appendix Table 12-3.) The descriptive measures contained in these tables were used to produce Figs 12-1 and 12-2, which graphically illustrate the growth and decline in the percentage of U.S. federal income tax returns with IRA contributions (1975–1996) and the growth in Canadian federal income tax returns with RRSP contributions (1968–1995), respectively.

For the tax years preceding 1982, any employee or self-employed U.S. taxpayer not covered by other deferred compensation plans could deduct up to $1,500 each year for contributions to an IRA. This maximum annual contribution limit increased to $2,000, beginning with the 1982 tax year.

If married, both spouses qualified, separately, under these rules. Therefore, each spouse could deduct a maximum IRA contribution of $2,000 per year, every year (1982–). IRA deduction or contribution ceilings were also limited to "compensation" or "earned income". Therefore, to qualify for a deduction for the maximum annual IRA contribution of $2,000, each taxpayer must have earned income of at least this amount. Alternatively, married taxpayers with

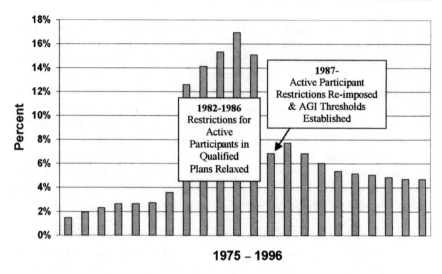

Fig. 12-1. U.S. Federal Income Tax Returns with IRA Contributions.

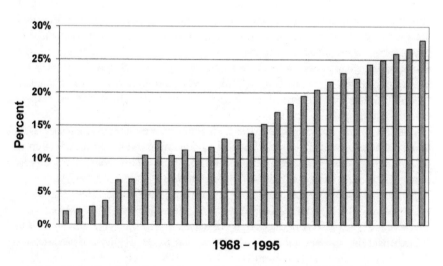

Fig. 12-2. Canadian Federal Income Tax Returns with RRSP Contributions.

unemployed spouses could only contribute an additional $250 per year (e.g. spousal IRA).

Beginning with the 1987 tax year, a phase-out was imposed for the deductible portion of IRA contributions for taxpayers with AGIs above "applicable dollar limits". For single or head of household (married, filing jointly or qualifying widow(er)) taxpayers with an AGI of $25,000 ($40,000) or less, the full $2,000 deduction remains available.[103] The deductibility of IRA contributions is fully phased-out for single of head of household (married, filing jointly or qualifying widow(er)) taxpayers at AGI levels of $35,000 ($50,000) or more. The maximum deduction is reduced on a five-to-one ratio. For every five dollars in additional AGI the maximum IRA deduction is reduced by one dollar. Though denied the tax benefits/savings associated with the IRA deduction, taxpayers above these filing status-based thresholds continue to be permitted to make *nondeductible* IRA contributions.

These IRA deduction phase-out thresholds begin, enter a flat range (like the EIC from Chapter 7) and remains constant, and are completely phased-out at the above levels for the 1987–1997 tax years, after which they are statutorily increased in varying increments for the different filing statuses. Projected increased thresholds exist beyond the 2007 tax year (e.g. the phase-out is scheduled to begin at an AGI of $80,000 for married taxpayers, filing a joint return, for the 2007 tax year), but, as is often the case, are subject to revision and/or modification to inflation-indexed measures.

Figure 12-1 graphically depicts U.S. taxpayer participation rates (1975–1996), in the form of percentages of total U.S. FIT returns filed with IRA contributions/deductions. During the 1982–1986 tax years, participation rates increased from less than 4% (1981) to more than 12%. This explosive increase appears to have been a function of the relaxed "active participants" restriction.

Prior to the 1982 tax year (and after the 1986 tax year), active participants in qualified plans (e.g. deferred compensation plans with their employers) were prevented from receiving tax benefits/savings from additional IRA contributions. For the post-1986 period, high-income taxpayers, operationally defined as those with AGIs above specified statutory thresholds, were permitted to make IRA contributions subject to phase-outs.

SOME STATISTICAL EVIDENCE

To statistically explore the influence of the "active participant" (AP) restriction (and other variables) on these U.S. IRA participation percentages, equation (12-1) was developed, as follows:

$$PERCENT = Intercept + AP + AGI + MAXFIT + Error \qquad (12\text{-}1)$$

where PERCENT, the dependent variable, equals the percent of total U.S. FIT returns filed with IRA contributions/deductions. The independent variables included AP, represented by dummy variables – the presence (1; 1975–1981 and 1987–1996) or absence (0; 1982–1986) of these restrictions for "active participants" in alternative deferred compensation plans, expected to reduce IRA participation; AGI phase-outs and restrictions are represented by dummy variables – the presence (1; 1987–) or absence (0; pre-1987) of these restrictions also inhibit IRA deductions/participation; and MAX FIT equals the maximum FIT rate, which declined from 70% (1975–1981) to 50% (1982–1986) to a blended rate of 38.5% (1987) to 33% (1988–1990) to 31% (1991 and 1992) to the contemporary maximum FIT rate of 39.6% (1993–1996). Higher (lower) FIT rates are expected to increase IRA participation rates.

The same independent variables were included in a second model, replacing PERCENT with DOLLARS as the dependent variable. This second model is represented by Eq. (12-2), as follows:

$$DOLLARS = Intercept + AP + AGI + MAXFIT + Error \qquad (12\text{-}2)$$

where DOLLARS equals the aggregate dollar amount of IRA contributions/ deductions. (Consumer price index-Urban (CPI-U) and maximum IRA contribution ceilings were also included in the preliminary analyses for both Eqs (12-1) and (12-2), but these variables were not significant and were removed from both equations.) The results of Eqs (12-1) and (12-2) are contained in Table 12-1.

For both Eqs (12-1) and (12-2), the C-p measure of model fit is favorable at 4.0. Also, for both Eqs (12-1) and (12-2), the AP independent variable is significant (S) and is in the hypothesized indirect direction. As active participants restrictions were relaxed, both percentage and dollar amount participation increased. Furthermore, the MAXTAX independent variable is in the hypothesized indirect direction, but is not significant (NS). Finally, the AGI independent variable produced insignificant and mixed coefficient signs for the two equations.

These two very simple equations produced intuitively appealing results and may provide the foundation for further investigation into the effects of additional or alternative variables. Clearly, any longitudinal research efforts into individual IRA participation rates or dollar (contribution) amounts must control for the significant impact of the reduced AP restrictions (1982–1986) period.

Table 12-1. Results of Regression Models for Equations (12-1) and (12-2) (t-values & F-statistics in parentheses).

| Independent Variables (IVs) | Dependent Variable (DV) – Regression Equations | | |
	PERCENT (12-1)	**DOLLARS** (12-2)	Relation
Intercept	0.2*** (4.4)	30.5** (3.5)	
AP Active Participant Restriction	–0.1*** (–5.1)	–30.2*** (–6.8)	Indirect & S As predicted
AGI AGI Limitation & Phase-Out	–0.0 (–0.2)	4.8 (0.7)	Mixed & NS
MAXTAX Maximum U.S. Marginal FIT Rates	–0.1 (–1.2)	–6.4 (–0.3)	Indirect & NS
N	22	22	
Adjusted R^2	94.9%	96.0%	
F for Regression	131.1***	170.7***	

*** $p < 0.001$
** $p < 0.01$

THE CANADIAN RRSP

The Canadian counterpart to the U.S. IRA is the registered retirement savings plan (RRSP). Unlike the U.S. IRA, the Canadian RRSP has never been subject to "active participant" restrictions, but is available to every Canadian citizen. Furthermore, the RRSP contribution is cumulative. For example, the Canadian taxpayer, unable to make a contribution for the 1995 tax year may accumulate this amount, to be carried forward, and contribute the balance available for the 1995 tax year at some later date. This option does not exist under the U.S. IRA counterpart.

The Canadian counterpart to the U.S. IRA, the RRSP, is graphically depicted (1968–1995) in Fig. 12-2. Unlike the U.S. IRA, the Canadian RRSP has shown a general trend of increasing participation. Like the U.S. IRA, the Canadian RRSP has an annual contribution limit, which excludes unused RRSP deduction accumulation from prior years.

SUMMARY

Chapter 11 provided coverage and discussion of the U.S. Social Security system. A rising stock market over the 1995–1999 period, combined with increased individual taxpayer interest in deferred compensation, led to the introduction and serious discussion of partial privatization of these social programs. However, these discussions preceded U.S. stock index declines experienced during the latter nine months of the 2000 calendar year.

A rising U.S. stock market has greatly contributed to increased individual taxpayer awareness and interest in the potential for deferred compensation growth in IRA, SEP, SIMPLE, and Keogh plans. Despite the very recent declining stock index values, interest in the privatization and individual control of some portion of Social Security contributions is likely to remain appealing to younger Americans. Therefore, we believe that U.S. tax policy will provide for (at least) partial Social Security contribution privatization in the next decade.

CHAPTER 13
ESTATE AND GIFT TAXATION

On September 8, 1916, there became effective a tax by the Federal Government on the transfer of estates of persons dying subsequent to that date . . . the taxes imposed by the prior acts were based upon the legacy or distributive share passing to the beneficiary and not upon the estate as a whole (IRS SOI, 1923, 33–34).

One primary impetus for imposing taxes upon the occasion of a person's death has been . . . the desire to reduce the degree of inequality in the distribution of wealth . . . (b)ut . . . also diminishes the incentive to accumulate capital to pass on to heirs (Wagner, 1977, 5).

Treasury officials . . . say only about 1.8% of Americans who die each year leave estates large enough to be taxed (WSJ, November 3, 1999, 1).

INTRODUCTION

Estate and gift taxes represent one comprehensive *excise tax* on the gratuitous transfers of property from individuals. In recent years, both Democrats and Republicans have introduced legislation for the partial or complete elimination of estate and gift taxation. Therefore, in the very near-term, we would anticipate a significant reduction – if not the complete elimination – of the already insignificant application of estate and gift taxation.

Generally, the *flat tax* and *national sales tax* proposals, discussed in Chapter 4, would suggest the complete repeal of estate and gift taxation. We do not anticipate this extreme change (or paradigm shift) in U.S. taxation.

Furthermore, the repeal of estate and gift taxation would also eliminate the step-up in tax basis, which is associated with estate and gift taxation. Therefore, the elimination of gift and estate taxation would result in the imposition of capital gains taxation on inherited property. Gains (and losses) would be calculated and based on the donor's cost of inherited property. Capital gains and losses were examined in Chapter 10.

A summary of some of the numerous estate and gift tax reform proposals, introduced in recent years, follows:

• H. R. 902/S. 75, introduced by Representative Christopher Cox (R-CA) and Senator John Kyl (R-AZ) and entitled "Family Heritage Preservation Act";

- H. R. 3076, introduced by Representative Max Sandlin (D-TX);
- H. R. 249, introduced by Representative Joseph Pitts (R-PA);
- H. R. 736, introduced by Representative Bob Stump (R-AZ);
- H. R. 1208, introduced by Representative Wes Watkins (R-OK);
- H. R. 525, introduced by Representative Philip Crane (R-IL);
- H. R. 802, introduced by Representative William Thornberry (R-TX); and
- S. 29, introduced by Senator Richard Lugar (R-IN), suggests outright repeal of the estate tax.
- H. R. 3879, introduced by Representatives Jennifer Dunn (R-WA) and John Tanner (D-TN) would phase out estate taxes over a ten-year period by reducing tax rates by five percentage points per year.
- S. 31, also introduced by Senator Richard Lugar (R-IN), would raise the unified credit over the next five years;
- H. R. 1584, introduced by Representative Sam Johnson (R-TX), provides a similar approach (raising the exclusion amount by $500,000 per year for five years) in the House; and
- H. R. 245, introduced by Representative Michael Pappas (R-NJ), also proposes a phase-out reform.
- S. 1711, introduced by Senator Kay Bailey Hutchinson (R-TX), would cut the top estate tax rate from 55% to 28%.
- S. 650, introduced by Senator Don Nickles (R-OK), provides for a tax rate of 20% on estates exceeding $1 million, and 30% on estates over $10 million.
- H. R. 8, recommending the 10-year phase-out/complete elimination of the "death tax" and entitled "Death Tax Elimination Act of 2000", was vetoed by President Clinton on August 31, 2000.

BACKGROUND AND DISCUSSION

In the U.S., the federal estate tax is a tax imposed on the *net estate* of a decedent. The net estate represents the value of property at the date of death. Gifts made by the *testator* in contemplation of death (i.e. inter vivos) are also considered part of the decedent's estate. Gift taxes are designed to prevent complete avoidance of the estate tax. The integration of estate and gift taxation has several advantages (Shoup, 1966, 14–15):

(1) It prevents the complete avoidance of the estate tax. This assumes a condition where the gift tax rate is below that provided for the estate tax, a situation when gifts are made in contemplation of death, or when the donor retains some interest in the property transferred;

(2) It provides for some degree of simplification (e.g. using similar tax rates schedules); and

(3) It encourages lifetime giving, reducing *lock-in* effects.

To be complete, the integration of estate and gift taxes requires (Shoup, 1966, 16–17):

(1) A *single* rate scale;
(2) A *single* applicable exemption amount;
(3) A *modest* annual (gift tax) exclusion per donee for each donor;
(4) Tax on lifetime gifts must be based on an amount *equal* to net gifts *plus* the gift tax paid; and
(5) The *same* inter vivos and testamentary methods for capital gain and loss taxation.

Beginning with the 1977 tax year, a single tax rate (the unified transfer tax) applied to estates and gifts. Beginning in 1982, the unlimited marital deduction became available. This made an estate of any size tax-free to a spouse.

SOME EARLY VIEWS OF ESTATE AND GIFT TAXATION

In his introduction to Eugenio Rignano's book, *The Social Significance of the Inheritance Tax*, Seligman (1924, 9) notes that

> . . . the demand for high inheritance taxes will continue to be an important plank in the platform . . . of socialists . . .

Rignano contended that property acquired and accumulated by the individual during his lifetime and own exertions should be free from taxation, but that the portion inherited from others should be subject to heavy taxation. His position was that these measures would strengthen (and not weaken) incentives to work, save, and accumulate wealth.

Alternatively, Sennholz (1976, 9) in *Death and Taxes*, a book published by the Heritage Foundation, took the reverse position

> A progressive estate tax that seizes and consumes productive capital . . . is a powerful instrument for inequality.

CONTEMPORARY ARGUMENTS FOR AND AGAINST ESTATE AND GIFT TAXATION

It is not difficult to identify the numerous organizations promoting the abolition of U.S. estate and gift taxation . . . or what has also been broadly referred to as the *death tax* (e.g. Newspaper Association of America (NAA), United Seniors Association, American Family Business Institute (AFBI), U.S. Chamber of Commerce, 60+, Family Business Tax Coalition, CATO Institute, Center for Study of Taxation, Citizens for a Sound Economy (CSE), The Heritage Foundation, Institute for Policy Innovation, National Federation of Independent Businesses (NFIB), Policy & Taxation Group, The Tax Foundation, and President George W. Bush). According to the NAA, the top ten reasons for the phase-out of the estate (and gift) tax include the following:

(1) It reduces the stock of capital in the economy;
(2) Ninety percent of family businesses fail shortly after the death of the founder;
(3) Ninety-one percent of all businesses in America are family owned;
(4) Collections represent less than 1.4% of total federal revenues; enforcement costs 65 cents for every $1 received;
(5) It is imposed on assets already taxed once or twice at the federal level;
(6) Family businesses divert significant amounts of economic resources to combat the estate tax (e.g. insurance and estate tax planning);
(7) U.S. estate tax rates exceed their counterparts in other industrialized nations;
(8) Repeal of the estate tax would reduce economic incentives to develop land and liquidate natural resources;
(9) It undermines the U.S. supported principles of hard work, savings and fairness; and
(10) Repeal of the estate tax is popular.

Alternatively, the Citizens for Tax Justice (CTJ) and 2000 Presidential Candidate, Vice-President Al Gore, suggested that estate and gift taxation should be retained:

(1) Fairness and equity is achieved through equal opportunity – a limitation on any gigantic head start. The estate tax reduces the concentration of wealth;
(2) The estate tax is extremely progressive and, therefore, perceived by many as fair or equitable. Revenues come from the wealthiest 1.4% – two-thirds from the wealthiest 0.2%. Without the estate tax, unrealized gains would remain forever untaxed;
(3) Sloth is discouraged. If providing a welfare mother $10,000 per year stifles work incentives, imagine a far greater windfall to a son or daughter;
(4) Charitable giving is encouraged. For 1995, 146 estates were responsible for about 40% of all reported charitable bequests;

(5) Fewer than 1 in 20 farmers leave a taxable estate, the typical tax payment for which is about $5,000. Non-farm family businesses represent less than 3% of total assets for estates worth less than $2.5 million (e.g. *small* businesses); and

(6) The $27 billion projected to be raised by the estate tax for fiscal 2000 is more than double the total amount of federal income taxes paid by the bottom half of all taxpayers. Any budget shortfall from the elimination of the estate tax would, for example, require a tripling of the taxes on these families.

Appendix Table 10-1 provided an international summary of the maximum capital gains tax rates for 100 countries as of January 1, 1991. This table also summarized maximum estate and gift tax rates. Figure 13-1 illustrates those countries with and without estate/gift taxation.

Fig. 13-1. Countries with and (without) Estate/Gift Taxation (shaded) – as of January 1, 1991.

The estate/gift tax results are comparable to those identified for capital gains tax rates (refer to Appendix Table 10–1). Approximately 36% of these countries did not have an estate tax. Approximately 38% of these countries did not have a gift tax. Alternatively, approximately 9% had a net worth tax.

U.S. ESTATE TAXATION

As first noted in Chapter 3, the estate tax (popularly referred to as the *Federal inheritance tax*)[104] was introduced with the *Revenue Act of 1916* (Sennholz, 1976, 1):

> Enacted in 1916, the federal estate tax laws included a specific exemption on the first $50,000; a minimum tax rate of 1% on the first $50,000 of taxable estates; and a maximum tax rates of 10% on taxable estates over $5 million ... (and) ... increased the taxes in 1917, 1918, 1921, 1932, 1934, 1935, and 1941.

Even in the early years of estate taxation, estate tax returns (Form 706) were not due until one year after death. Estate tax returns were (and remain) subject to the statutory provisions in effect at the date of death.

Initially, the estate tax imposed a maximum progressive tax rate/bracket of 10% on the transfer of the net estates (of taxable size) of persons dying on or after September 9, 1916–March 2, 1917. Presently, a minimum (maximum) estate tax rate of 18 (55)% applies and unrealized capital appreciation is included and subject to estate tax.

Appendix Table 13-1 provides a summary of minimum and maximum estate tax rates, the number of estate tax returns filed, and the basic exemption[105] or contemporary unified[106] credit equivalent for the entire history of the post-16th Amendment estate tax.

Generally, the largest components of assets reported in both taxable and nontaxable U.S. estate tax returns are represented by marketable equity securities (i.e. stocks), real estate, and debt securities (i.e. bonds), in descending order. On average, these items represent approximately 70% of the value of gross assets contained in the contemporary estate (IRS SOI, 1996–1997, 21).

U.S. GIFT TAXATION

As first noted in Chapter 3, the gift tax was introduced with the *Revenue Act of 1924*. It was imposed on donative transfers beginning with the 1924 calendar year and used the same tax rates used for the estate tax. No similar tax had ever been imposed by any prior act of Congress (IRS SOI, 1924, 96).

The gift tax (at a maximum rate of 25%) was repealed by the *Revenue Act of 1926* for gifts made after January 1, 1926. The *Revenue Act of 1932* re-imposed the gift tax for gifts made after June 6, 1932, at a maximum rate of 33.5% (IRS SOI, 1932, 216).

Appendix Table 13-2 provides an historical (1932–1951) summary of the number of gift tax returns filed, the percent taxable, net taxable gift amounts, and the dollar amounts of the gift tax.

CANADIAN ESTATE (SUCCESSION) AND GIFT TAXES

> The Canadian provincial (succession duties) legislation was inspired by and modeled) after legislation of the American states. Clear testimony to this fact is given in the work of R. A. Bayly (1902) . . . "That the New York and Pennsylvania Acts in force at that time were copied in principle and in detail must be at once apparent to anyone who compares them with the Ontario Act of 1892 . . . (t)he Ontario Act of 1892 was purely American in its origin" (Perry, 1955, 110–11).
> Succession duties and gift taxes were . . . significant . . . for both the federal and provincial governments during the 30 years prior to the federal elimination of these taxes at the end of 1971. By the end of 1979, all provinces but Quebec had withdrawn from the estate and gift tax field. Quebec eliminated taxes on succession and gifts made after April 23, 1985 (Treff & Cook, 1995, 3:28).

Canada's Dominion Succession Duty Act, establishing the Canadian estate tax, became effective and applied to the estates of decedents dying on or after June 14, 1941. The Estate Tax Act was established and applied to taxpayers dying on or after January 1, 1959. The Canadian federal estate tax was eliminated on December 31, 1971.

The former established five separate classes of beneficiaries and rates schedules, from the most-favored to the least-favored, as follows (DNR, 1946, 124–125):

- Class A: Wife; children under 18; children over 18 if dependent on the deceased due to mental or physical disability.
- Class B: Husband; parents; grandparents; son or daughter over 18; son or daughter in law.
- Class C: Brother or sister; uncle or aunt; cousins and any decedents of these.
- Class D: Other distant relatives; non-blood relations; charitable bequests in excess of 50% of the aggregate estate.
- Charity: Exempted charities, as defined by the ITA, but limited to 50% of the aggregate estate.

All estates with an aggregate net value of less than $5,000 were exempt. Tax rates ranged from 0%–27%. Appendix Tables 13-3 and 13-4 provide

UNITED STATES (1916–)

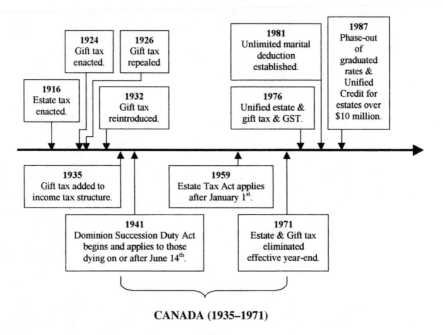

CANADA (1935–1971)

Fig. 13-2. A Summary of Effective Dates of U.S. and Canadian Estate and Gift Taxation.

summaries of the descriptive statistics relating to Canadian estate and gift tax returns, respectively.

The Canadian Federal gift tax was eliminated for gifts made after December 31, 1971. Figure 13-2 provides a timeline of the evolution of U.S. and Canadian estate and gift taxation. Estate taxation began in the U.S. in 1916. Estate and gift taxation still exists in the U.S. For Canada, gift and succession taxation was in effect for 1935–1971.

SUMMARY

Nearly a quarter of a century ago, Wagner (1977, 23) suggested that transfer (e.g. estate and gift) taxation was a relatively unimportant source of revenue, but represented an egalitarian notion of promoting *equality* in the distribution of wealth. Some view estate and gift taxation as a mechanism for the reduction of inequalities normally (and cumulatively) evolving in a market economy.

Estate and gift tax returns are projected to generate 137,000 and 292,000 tax returns for the 2000 calendar year, respectively. As previously noted, estate and gift taxation generates relatively little tax revenue, is complex, and administration and collection costs are high. Therefore, the only valid reason for the maintenance of estate and gift taxation would appear to relate to the U.S. public's *perception* of fairness and equity.

Few in the popular press have focused on the fact that the elimination of the estate tax, and the related *step up* in tax basis for capital assets transferred to heirs, would broaden the applicability of undesirable *lock in* effects. Decreases in tax revenues from estate and gift taxes would be partially offset by increased capital gains tax revenues.

The U.S. did not eliminate estate and gift taxation in the early 1970s, when Canada did (see Fig. 13-2). We do not expect that it will be completely eliminated in the early decades of the new millenium. Because of the close race for the Presidency in November/December 2000, we anticipate a hybrid of the President Bush and Vice-President Gore proposals and a reduction, but not the complete elimination, of estate and gift taxation. Since proposals by both political parties would exempt most family farms and businesses from estate and gift taxation, we anticipate the early passage of legislation favoring these groups. However, a short-lived economic downturn, reduced U.S. budget surpluses or a return of Federal budget deficits, or even a short-lived stock market decline (or decreasing stock price increases) would set the political stage for the complete or partial restoration of any contemporary changes to estate and gift taxation.

PART 5

PROSPECTIVE INDIVIDUAL TAX POLICY ISSUES

CHAPTER 14
TRENDS IN INDIVIDUAL TAXATION

Canadians chose a Liberal plan that is based on the fundamental belief that the surplus belongs to all citizens . . . the Reform Alliance's ultimate objective is to force "all workers to invest in and manage more of their own retirement savings" by imposing "mandatory private savings plans" (http://www.liberal.ca, December 14, 2000, regarding the November 27, 2000, election of Jean Chrétien as the new Canadian Prime Minister).

INTRODUCTION

Many of the tax policies appearing in the last quarter of a century appear to represent U.S. variations of Canadian-initiated measures:

First, the U.S. is increasingly using tax *credits*. In the past, increases in tax *deductions* would have been a more likely tax policy mechanism. This may be due, in no small part, to the great success of the bi-partisan passage (1975–) of the earned income tax credit (Chapter 7). However, the Canadian objective for replacing the personal exemption *deduction* (Chapter 6) with a tax *credit* (1987–), at the same/lowest tax bracket for all taxpayers, was promoted and continues to be perceived (by Canadians) as preferable and more equitable. This position is inconsistent with the historical and contemporary U.S. position that progressive taxation is more equitable than a flat tax or a national sales tax (Chapters 2–5). It is, however, important to note that an important contemporary U.S. policy issue is the elimination of the so-called "marriage tax" (Chapter 8). The Canadian system of taxing the *individual*, instead of the *household*, eliminates this possible penalty.

Second, the use of tax deferred IRA-like accounts is expanding. This is due to the overwhelming success of deferred compensation plans (Chapter 12). The availability of the Canadian RRSP (1958–) preceded that of the U.S. IRA (1975–). Furthermore, in both U.S. and Canadian (November 2000) elections, proposals for the initiation of taxpayer management of their own Social Security (U.S.) or Canadian Pension Plan (Canada) contributions were proposed (Chapter 11). In the U.S., this proposal was included in President

155

Bush's plan. In Canada, a similar proposal was included in the conservative Reform Alliance's political platform.

Third, the U.S. in now considering a reduction (or the complete elimination) of estate and gift taxation (Chapter 13). In the U.S., the imposition of estate (1916–) and gift (1924–) taxes preceded similar tax policy measures in Canada. Canada adopted the gift tax in 1935 and the counterpart to the U.S. estate tax in 1941. Canada eliminated their estate and gift taxes in 1971.

Alternatively, both Canada and the U.S. continue to retain some form of alternative minimum tax (Chapter 9). In this case, the Canadian "minimum tax" (1986–) was introduced *after* the contemporary U.S. version (1979–). Both nations also continue to tax capital gains (Chapter 10).

Furthermore, the U.S. has, thus far, rejected other Canadian policies, including: (1) the imposition of a national sales tax (Chapter 4), (2) the socialization or establishment of a national health care program, and (3) a national retirement savings plan that is financed *directly* from general revenues (Chapter 11).

The *progressive* system of taxation appears destined to remain a stable foundation for the evolution of future U.S. tax policy. Similarly, the *household* appears likely to remain the relevant unit of taxation. U.S. tax policy is likely to continue to selectively employ many of Canada's more palatable initiatives.

THE 2000 PRESIDENTIAL ELECTION

As this monograph goes to press, we have just concluded a bitterly extended Presidential race between Vice-President Al Gore (D) and, now, President George W. Bush (R). The views of these candidates/parties, as they relate to this monograph, are summarized in Table 14-1.

SUMMARY

As a candidate, President Bush and/or the Republican party wanted to return at least 25% of anticipated Federal budget surpluses to U.S. taxpayers in the form of lower marginal tax rates. Their concern is that any failure to do so would tempt government officials to expand the Federal Government instead of reducing the National debt. Mechanisms for achieving this objective include the restoration of the $3,000 special deduction for two-earner couples to reduce the so-called "marriage tax" (Chapter 8 and Appendix 3H); the reduction of the alternative minimum tax (Chapter 9), for married taxpayers; and the ten-year phase-out of the estate and gift or "death" tax, (Chapter 13). President Bush

***Table 14-1*ª.** 2000 Presidential Election Year Tax Policy Issues.

Chapter & Tax Policy Issue	Al Gore & Democrats	G. W. Bush & Republicans
Chapter 2: Contemporary Income Tax	No basic rates change.	Reduce all tax rates: from 15% to 10%; 28%/31% to 25%; & 36%/39.6% to 33%.
Appendix 3C: Charitable Contributions	No change.	Expand applicability of the charitable contribution deduction to non-itemizers.
Chapter 4: Progressive Tax Rates	No change.	Reduce the rate of progressivity.
Chapter 6: Child Tax Credits Standard Deductions	Increase. Double for married	Double to $1,000. No change. taxpayers to reduce the marriage tax penalty.
Chapter 7: Earned Income Tax Credit	Increased benefits before the phase-out applies.	Eliminate the EITC component of the "marriage tax".
Chapter 8: Marriage Tax Penalty	Double the standard deduction for the single	Restore the special $3,000 two-earner deduction. taxpayer.
Chapter 9: Alternative Minimum Tax	No change.	Eliminate the AMT component of the marriage tax penalty.
Chapter 10: Capital Gains	No change.	No change.
Chapter 11: Social Security	Devote all Social Security surpluses to debt reduction and Social Security only; expand Medicare to include a prescription drug benefit.	Voluntary portion (e.g. 15%) of Social Security to be under taxpayer control; expand Medicare to include a prescription drug benefit.
Chapter 12: Deferred Compensation	Create 401(J) Life-Long Learning accounts and USA accounts.	See Social Security.
Chapter 13: Estate and Gift Taxation	Increase exemptions.	Completely eliminate.

ª Much of this information was developed from candidate Web sites at http://www.algore2000.com and http://www.georgebush.com on July 22, 2000.

also proposed taxpayer control for a portion of their Social Security payroll taxes (Chapter 11).

Candidate Vice-President Al Gore and the Democrats are opposed to the complete elimination of all components of the "marriage tax", restricting relief to the base effect (or the standard deduction component), which only applies to non-itemizer taxpayers (Chapter 8). They also opposed the complete elimination of estate and gift taxation (Chapter 13).

Both candidates and/or parties are promoting expansion of Medicare to include prescription drugs and increases in the child tax credit (1997–).

Alan Greenspan, the Chairman of the Federal Reserve, previously cautioned that any *transitory* or *non-permanent* budget surpluses, the component of which has not yet been determined, should not lead politicians to legislate *permanent* tax reductions. However, as this monograph goes to press, we anticipate (1) a reduction or elimination of the marriage tax (Chapter 8), (2) a reduction or elimination of estate and gift tax rates or applicability (Chapter 13), and (3) the reduction of existing progressive tax rates (Chapter 4).

CHAPTER 15
CONCLUSIVE REMARKS

The economical systems of a nation are the result of growth. Changes should be made slowly. The fact that a system of taxation has many imperfections is no argument in favor of radical or hasty reforms. It would be rash in the extreme to abolish all the old forms of taxation, and adopt in their stead new and untried theories. Great harm would be done before a country could adjust itself to the new order of things (Burke, 1891, 249).

Old programs with outdated aims stay in place. Newer ones, added piecemeal, often conflict with the old . . . rather than adopt wholesale reform, the lawmakers try to give relief through tiny, symbolic projects (Shlaes, 1999, 21 and 25).

INTRODUCTION

From a historical perspective, the contemporary U.S. individual federal income tax remains relatively flat (Chapter 3). *Apparent* progressivity has been replaced with *effective* progressivity, through mechanisms like standard deduction and personal exemption phase-outs (Chapter 2 and Appendix 2A). Similarly, the alternative minimum tax has evolved to ensure equity with respect to tax preferences for high-income taxpayers (Chapter 5 and Chapter 9), and limitations on the deductibility of interest deductions (Appendix 3G) and passive activity losses (Appendix 3I) prevent high-income taxpayers from taking advantage of previously available tax avoidance mechanisms. Flat tax, national sales tax, and even Internet sales tax proposals have been examined over the past decade, but these have not generated broad support (Chapter 4 and Appendix 4K).

We expect the *progressive* nature of the U.S. system of taxation to continue (at least) through the early decades of the new millenium. Furthermore, we expect the implementation of a national sales tax (or the replacement of our progressive system of taxation with a flat/national sales tax) to continue to be perceived as regressive or inequitable (and even socialistic) and to remain politically infeasible. We do, however, anticipate the application of some form of (non-national) Internet sales taxation.

Capital gains taxation (Chapter 9) was a non-issue during the 2000 Presidential campaign. We expect capital gains to continue to be taxed as long as U.S. inflation rates remain low (Appendix 5L) and the recent historical U.S. budget surpluses continue to be associated with the (transitory?) gains experienced through the April 2000 "crash" (Appendix 10N). A revival of the "bull" (stock) market gains experienced through April 2000 would likely result in broad support for the privatization of some component of Social Security contributions (Chapter 11).

A reduction (or the complete elimination) of the so-called "marriage tax" (Chapter 8 and Appendix 3H) represents a renewed desire to identify more effective tax policy alternatives to the failed social service mechanisms of the past. The objective is to assist in the effort to preserve the family as the fundamental U.S. "institution" through additional child care credits (Chapter 6) and the expansion of the earned income tax credit (Chapter 7). The EIC (1975–) represents one of the most successful tax policy measures ever implemented.

The wealth created by the Internet and technology has led to a general pattern toward privatization. The "portability" or "privatization" of previously government-controlled social programs is, perhaps, most evident in the case of the IRA or deferred compensation plans (Chapter 12). Health insurance/care appears to be evolving in a similar manner, as is education and long-term disability. The pattern appears to remain toward provision of social services, but through tax credits, retaining the "privatization" or "market-based" selection system that is fundamental to capitalism. Energy independence proposals are consistent with this pattern, as is a reduction (or complete elimination) of estate and gift taxation (Chapter 13). These and related matters are discussed below.

Health Insurance

Health care remains an issue of concern to most Americans. Anecdotal evidence suggests that the Canadian system of free national health care is ailing. Canadians travel to the U.S. for necessary operations that the Canadian health care system cannot provide on a timely basis. However, the cost of prescription drugs (not covered by the Canadian health care system) is significantly lower in Canada when compared to the U.S.

Health care and prescription drugs plans were prominent in the 2000 Presidential campaign debates. Candidate Vice-President Al Gore proposed a mechanism to subsidize the purchase of private insurance by individual taxpayers. His proposal was a 25% tax credit for individual taxpayers without employer-provided health insurance. President G. W. Bush also proposed a limited tax credit, in addition to an expansion of the Medical Savings Account.

For tax years beginning after 1996, the *Health Insurance Portability and Accountability Act* (PL 104–191; popularly known as the Kassebaum-Kennedy Bill) provided for the creation of the Medical Savings Account (MSA). MSA provisions created section 220 of the Internal Revenue Code (IRC).

The MSA "pilot project" provided for a four-year trial period (1997 through 2000), and was created to benefit self-employed individuals and employees of small employers to meet the costs of medical care.[107] At the end of this trial period, Congress will vote before the bill is expanded to cover more than the 750,000 policy ceiling originally provided for. However, as of June 30, 1997, only about 17 thousand MSAs had been established (RIA, 1997, 210).

The MSA is an IRA-like banking account, established with a financial institution or insurance company. (IRAs and other deferred compensation programs were discussed in Chapter 12.) Contributions are tax-deductible (above-the-line). (The general framework of above-the-line and itemized deductions are illustrated in Appendix 2A.) Distributions for qualified medical expenses are not taxable. Unspent contributions continue to build, tax-free, for future medical expenses or retirement (at age 65).[108]

The conditions required to qualify for the MSA include: (1) employment with a "small employer"[109] or self-employment and (2) participation in a "high-deductible"[110] health care plan.

The focus of the MSA on the small employer and self-employed taxpayers is a natural progression, not unlike the Social Security system. Recall that the Social Security system first covered "employees" in 1937 and followed with coverage for the "self-employed" in 1951 (see Chapter 11).

Unlike Canadian health care, the U.S. MSA retains the "free-enterprise" or "profit-motivated" feature of the existing U.S. health care system. It achieves this objective within the framework of a tax-deferred IRA-like mechanism (see Chapter 12). We believe that the U.S. MSA is preferable to a system representative of the ailing Canadian health care system. Continued discussion and further research into modifications to expand this program warrant investigation by academic researchers.

Education

A primary issue during the 2000 Presidential campaign was education. President Bush proposed and expansion of existing education IRAs. In addition, the Bush plan includes the use of a voucher system, to integrate certain elements of "market efficiency" or "privatization" into the public education system. Vice-President Gore proposed an increase in the value of and expanded eligibility for college tuition credits, as well as the creation of 401(j) Lifelong Learning accounts. Note that both proposals involve extensions of the "individual control" enjoyed by successful IRA and deferred compensation (Chapter 12) plans. We expect these (and additional) plans to enjoy increased support in the early decades of the new millenium.

Long-Term Care and Disability

During the 2000 Presidential campaign, candidate Vice-President Al Gore proposed: (1) a new tax credit for up to $3,000 for long-term care expenses, and (2) a new tax credit for up to $1,000 for work-related expenses for disabled workers. President G. W. Bush, as noted in the summary to Chapter 6, proposed an additional personal exemption for caregivers (of long-term care) and a deduction for long-term care insurance premiums. President Bush did not have a counterpart to the Gore proposal for disabled workers.

We expect that these measures, like the IRA and MSA, will take the form of above-the-line *tax deductions*. Tax policies favoring the deductibility of long-term care and disability expenses will remain politically popular as the U.S. population continues to age.

Energy-Efficiency

During the latter part of the 2000 calendar year, petroleum and energy costs rose significantly. This increase, though not resulting in crippling supply shortages, were reminiscent of the mid-1970s and illustrated the vulnerability of the U.S. to price-fixing and anti-competitive supply reductions by the OPEC oil cartel.

President Bush campaigned on increased exploration and the exploitation of U.S. petroleum reserves (e.g. Alaska), to lessen U.S. dependence on foreign oil. G. W. Bush has experience in the oil industry. Vice-President Gore proposed new tax credits for consumers who buy energy-efficient cars and new homes. We anticipate the near-term passage of hybrid legislation on both supplier and consumer fronts. Though beyond the scope of this monograph, tax credits and measures similar to those evolving during the petroleum shortages of the mid-1970s are likely to be revived.

SUMMARY OF EMERGING TAX POLICY ISSUES

We expect to see the greatest growth (or modification) in the early decades of the second millenium to take the form of *direct* taxation. We anticipate a focus on income and wealth equalization, but through "portability" (e.g. education, health care and disability insurance) and "privatization" (e.g. education, deferred compensation and Social Security retirement benefits). Social programs will continue to be Government funded, but through tax deductions, tax credits, and "individual control", with the choices available only in a "free-market" environment.

This monograph has examined historical and contemporary trends in U.S. individual taxation. To the extent practicable, we have attempted to develop these historical and contemporary patterns, and to suggest the future directions of tax policy likely to warrant further study.

We believe that the events that have led to this pattern of a shift from a "direct" to an "indirect" provision of social services has evolved from the tremendous success of: (1) the earned income credit (1975-), (2) the popularity (and profitability) of IRA and deferred compensation plans, (3) the "bull" stock market of the 1990s, and (4) very recent (and anticipated) U.S. budget surpluses. In the event of a stable or growing U.S. economy, we expect this pattern to continue.

APPENDIX I

(Supplemental Figures and Tables to Chapters 1–15)

Appendix Table 2-1. Phase-Out[1] of Personal Exemptions – by Filing Status (1991–2000).

Tax Year	Relevant AGI Threshold	
	Begin Phase-Out	End/Complete Phase-Out

Single (SGL):

Tax Year	Begin Phase-Out	End/Complete Phase-Out
1991	$100,000	$222,500
1992	$105,250	$227,750
1993	$108,450	$230,950
1994	$111,800	$234,300
1995	$114,700	$237,200
1996	$117,950	$240,450
1997	$121,200	$243,700
1998	$124,500	$247,000
1999	$126,600	$249,100
2000	$128,950	$251,450

Married, Filing Jointly (MFJ) or Surviving Spouse (SS):

Tax Year	Begin Phase-Out	End/Complete Phase-Out
1991	$150,000	$272,500
1992	$157,900	$280,400
1993	$162,700	$285,200
1994	$167,700	$290,200
1995	$172,050	$294,550
1996	$176,950	$299,450
1997	$181,800	$304,300
1998	$186,800	$309,300
1999	$189,950	$312,450
2000	$193,400	$315,900

Married, Filing Separately (MFS):

Tax Year	Begin Phase-Out	End/Complete Phase-Out
1991	$ 75,000	$136,250
1992	$ 78,950	$140,200
1993	$ 81,350	$142,600
1994	$ 83,850	$145,100
1995	$ 86,025	$147,275
1996	$ 88,475	$149,725
1997	$ 90,900	$152,150
1998	$ 93,400	$154,650
1999	$ 94,975	$156,225
2000	$ 96,700	$157,950

Appendix Table 2-1. Continued.

	Relevant AGI Threshold	
Tax Year	Begin Phase-Out	End/Complete Phase-Out
Head of Household (HH):		
1991	$125,000	$247,500
1992	$131,550	$254,050
1993	$135,600	$258,100
1994	$139,750	$262,250
1995	$143,350	$265,850
1996	$147,450	$269,950
1997	$151,500	$274,000
1998	$155,650	$278,150
1999	$158,300	$280,800
2000	$161,150	$283,650

[1] The personal exemption is reduced by 2% for each $2,500 ($1,250 for married taxpayers filing separately) or fraction thereof for taxpayers with an AGI in excess of the threshold amounts.

Appendix Table 3-1. Summary of U.S. Total, Taxable, and Percent Taxable Individual FIT Returns Filed (1913–1996).

Tax Year	Total Filed Individual FIT Returns	Less: Non-taxable FIT Returns	Equals: Taxable FIT Returns	Percent Taxable	Source: *IRS SOI*
For net income of $3,000 or more for the 1913 (for the ten months from March 1–December 31)–1916 tax years:					
1913	357,598	Unknown[1]	357,598	Unknown	1916, 14
1914	357,515	Unknown[1]	357,515	Unknown	1916, 14
1915	336,652	Unknown[1]	336,652	Unknown	1916, 14
1916	437,036	74,194	362,842	83.02%	1916, 26–7
For net income of $1,000 or more for the 1917–1921 tax years:					
1917	3,472,890[2]	732,841[3]	2,740,049	78.90%	1917, 10, 12 & 26–7
1918	4,425,114	1,032,251	3,392,863	76.67%	1918, 34–5
1919	5,332,760	1,101,579	4,231,181	79.34%	1919, 41–2
1920	7,259,944	1,741,634	5,518,310	76.01%	1920, 46–7
For all positive net income for the tax years 1922 forward:					
1921	6,662,176	3,072,191	3,589,985	53.89%	1921, 40–1
1922	6,787,481	3,106,232	3,681,249	54.24%	1922, 80
1923	7,698,321	3,428,200	4,270,121	55.47%	1923, 59
1924	7,369,788	2,880,090	4,489,698	60.92%	1924, 104
1925	4,171,051	1,669,885	2,501,166	59.96%	1925, 87
1926	4,138,092	1,667,102	2,470,990	59.71%	1926, 3
1927	4,101,547	1,660,606	2,440,941	59.51%	1927, 3
1928	4,070,851	1,547,788	2,523,063	61.98%	1928, 3
1929	4,044,327	1,586,278	2,458,049	60.78%	1929, 4
1930	3,707,509	1,669,864	2,037,645	54.96%	1930, 4
1931	3,225,924	1,700,378	1,525,546	47.29%	1931, 4
1932	3,877,430	1,941,335	1,936,095	49.93%	1932, 5
1933	3,723,558	1,975,818	1,747,740	46.94%	1933, 5
1934	4,094,420	2,298,500	1,795,920	43.86%	1934, 5
1935	4,575,012	2,464,122	2,110,890	46.14%	1935, 4
1936	5,413,499	2,552,391	2,861,108	52.85%	1936, 4
1937	6,350,148	2,978,705	3,371,443	53.09%	1937, 6
1937 and future years restated to exclude fiduciary/estate & trust returns (Forms 1041)[4]:					
1937	6,301,833	2,974,921	3,326,912	52.79%	1937, 35
1938	6,251,009	3,255,319	2,995,690	47.92%	1938, 131
1939	7,652,781	3,756,346	3,896,435	50.92%	1939, 131
1940	14,710,771	7,273,464	7,437,307	50.56%	1940, 145
1941	25,869,917	8,367,033	17,502,884	67.66%	1941, 151

Appendix Table 3-1. Continued.

Tax Year	Total Filed Individual FIT Returns	Less: Nontaxable FIT Returns	Equals: Taxable FIT Returns	Percent Taxable	Source: *IRS SOI*
1942	36,619,246	8,982,195	27,637,051	75.47%	1942, 161
1943	43,722,038	3,481,901	40,240,137	92.04%	1943, 33
1944	47,111,495	4,757,027	42,354,468	89.90%	1944, 36
1945	49,932,783	7,282,281	42,650,502	85.42%	1945, 36
1946	52,816,547	14,900,851	37,915,696	71.79%	1946, 38
1947	55,099,008	13,520,484	41,578,524	75.46%	1947, 37
1948	52,072,006	15,660,758	36,411,248	69.92%	1948, 36
1949	51,814,124	16,185,829	35,628,295	68.76%	1949, 46
1950	53,060,098	14,873,416	38,186,682	71.97%	1950, 39
1951	55,447,009	12,798,399	42,648,610	76.92%	1951, 28
1952	56,528,817	12,652,544	43,876,273	77.62%	1952, 21
1953	57,838,184	12,615,033	45,223,151	78.19%	1953, 26
1954	56,747,008	14,113,948	42,633,060	75.13%	1954, 36
1955	58,250,188	13,561,123	44,689,065	76.72%	1955, 21
1956	59,197,004	12,938,358	46,258,646	78.14%	1956, 3
1957	59,825,121	12,959,806	46,865,315	78.34%	1957, 3
1958	59,085,182	13,433,048	45,652,134	77.26%	1958, 3
1959	60,271,297	12,774,384	47,496,913	78.81%	1959, 3
1960	61,027,931	12,966,946	48,060,985	78.75%	1960, 4
1961	61,499,420	12,916,655	48,582,765	79.00%	1961, 3
1962	62,712,386	12,620,023	50,092,363	79.88%	1962, 4
1963	63,943,236	12,620,015	51,323,221	80.26%	1963, 4
1964	65,375,601	14,069,263	51,306,338	78.48%	1964, 2
1965	67,596,300	13,895,506	53,700,794	79.44%	1965, 1
1966	70,160,425	13,451,349	56,709,076	80.83%	1966, 2
1967	71,651,909	12,978,971	58,672,938	81.89%	1967, 2
1968	73,728,708	12,440,000	61,288,708	83.13%	1968, 2
1969	75,834,388	12,112,994	63,721,394	84.03%	1969, 2
1970	74,279,831	14,962,460	59,317,371	79.86%	1970, 2
1971	74,576,407	14,660,035	59,916,372	80.34%	1971, 8
1972	77,572,730	16,703,713	60,869,017	78.47%	1972, 8
1973	80,692,587	16,425,425	64,267,162	79.64%	1973, 7
1974	83,340,190	16,005,423	67,334,767	80.80%	1974, 9
1975	82,229,332	20,738,595	61,490,737	74.78%	1975, 13
1976	84,670,389	20,249,022	64,421,367	76.08%	1976, 2
1977	86,634,640	22,253,502	64,381,138	74.31%	1977, 1
1978	89,771,551	21,083,246	68,688,305	76.51%	1978, 1
1979	92,694,302	20,999,319	71,694,983	77.35%	1979, 2
1980	93,902,469	19,996,225	73,906,244	78.71%	1980, 2
1981	95,396,123	18,671,399	76,724,724	80.43%	1981, 2
1982	95,337,432	18,302,132	77,035,300	80.80%	1982, 2

Appendix Table 3-1. Continued.

Tax Year	Total Filed Individual FIT Returns	Less: Nontaxable FIT Returns	Equals: Taxable FIT Returns	Percent Taxable	Source: *IRS SOI*
1983	96,321,310	18,304,987	78,016,323	81.00%	1983, 2
1984	99,438,708	17,799,199	81,639,509	82.10%	1984, 10
1985	101,660,287	18,813,867	82,846,420	81.49%	1985, 12
1986	103,045,170	19,077,757	83,967,413	81.49%	1986, 16
1987	106,996,270	20,272,474	86,723,796	81.05%	1987, 20
1988	109,708,280	22,572,948	87,135,332	79.42%	1988, 18
1989	112,135,673	22,957,318	89,178,355	79.53%	1989, 18
1990	113,717,138	23,854,704	89,862,434	79.02%	1990, 16
1991	114,730,123	25,996,536	88,733,587	77.34%	1991, 20
1992	113,604,503	26,872,557	86,731,946	76.35%	1992, 31
1993	114,601,819	28,166,452	86,435,367	75.42%	1993, 31
1994	115,943,131	28,323,684	87,619,446	75.57%	1994, 38
1995	118,218,327	28,965,338	89,252,989	75.50%	1995, 35
1996	120,351,208	29,421,858	90,929,350	75.55%	1996, 35

[1] No detail of non-taxable returns is available for the 1913, 1914, or 1915 tax years.

[2] A total of 1,640,758 (47.2%) of the individual FIT returns filed had net income ranging from $1,000 to $2,000 (IRS SOI 1917, 5):

Intensive study was made only with respect to those reporting income of $2,000 or more.

A total of 1,832,132 (52.8%) of the individual FIT returns filed had net incomes above $2,000.

[3] The IRS SOI provided non-taxable return measures for returns with net incomes above $2,000. The analysis of 1918 and 1919 tax returns filed with net incomes ranging from $1,000 to $2,000 resulted in 32% and 28% non-taxable returns, respectively. Therefore, the 1,640,758 individual tax returns with net incomes ranging from $1,000 to $2,000 are presumed to have contained approximately 30% (the mean of 32% and 28%) non-taxable returns for an additional 492,227 non-taxable returns and a total estimate of 732,841 non-taxable returns for the 1917 tax year.

[4] For 1913–1937 data is comparable (IRS SOI 1937 (Part I), 1–2 & 35):

For the income year 1937, income from an estate or trust, whether or not taxable to the fiduciary, is required to be reported on the fiduciary income tax return, Form 1041. Prior to 1937, income from an estate or trust which was taxable to the fiduciary was required to be reported on the individual income tax return, Form 1040, and income from an estate or trust which was non-taxable to the fiduciary was required to be reported on the fiduciary income tax return, Form 1041, which was filed for information purposes.

Appendix Table 3-2. Summary of Canadian Total, Taxable, and Percent Taxable Individual FIT Returns Filed (1917–1995).

Tax Year	Total Filed Individual FIT Returns	Less: Non-taxable FIT Returns	Equals: Taxable FIT Returns[1]	Percent Taxable	Source: RC TS
1917	Unknown	Unknown	Unknown	Unknown	None
1918	Unknown	Unknown	Unknown	Unknown	None
1919	Unknown	Unknown	Unknown	Unknown	None
1920	Unknown	Unknown	Unknown	Unknown	None
1921	Unknown	Unknown	190,561	Unknown	1936, 3
1922	Unknown	Unknown	290,584	Unknown	1936, 3
1923	Unknown	Unknown	281,182	Unknown	1936, 3
1924	Unknown	Unknown	239,036	Unknown	1936, 3
1925	Unknown	Unknown	225,514	Unknown	1936, 3
1926	Unknown	Unknown	209,539	Unknown	1936, 3
1927	Unknown	Unknown	116,029	Unknown	1936, 3
1928	Unknown	Unknown	122,026	Unknown	1936, 3
1929	Unknown	Unknown	129,663	Unknown	1936, 3
1930	Unknown	Unknown	142,154	Unknown	1936, 3
1931	Unknown	Unknown	143,601	Unknown	1936, 3
1932	Unknown	Unknown	133,621	Unknown	1936, 3
1933	Unknown	Unknown	166,972	Unknown	1936, 3
1934	Unknown	Unknown	203,957	Unknown	1936, 3
1935	189,748	5,553	184,195	97.07%	1943, 5 & 1940, 8
1936	212,112	13,010	199,102	93.87%	1943, 5
1937	240,956	23,907	217,049	90.08%	1943, 5
1938	245,134	8,070	237,064	96.71%	1943, 5
1939	Unknown	Unknown	257,186	Unknown	1943, 5
1940	Unknown	Unknown	684,359	Unknown	1943, 5
1941	Unknown	Unknown	896,947	Unknown	1943, 5
1942	Unknown	Unknown	1,781,244	Unknown	1964, 88
1943	Unknown	Unknown	2,163,354	Unknown	1965, 100
1944	Unknown	Unknown	2,254,319	Unknown	1965, 100
1945	Unknown	Unknown	2,254,246	Unknown	1965, 100
1946	3,162,032	808,910	2,353,122	74.42%	1967, 104
1947	3,528,776	1,162,320	2,366,456	67.06%	1949, 109
1948	3,662,030	972,100	2,689,930	73.45%	1950, 113
1949	3,754,760	1,522,790	2,231,970	59.44%	1951, 111
1950	3,866,160	1,491,920	2,374,240	61.41%	1980, 174
1951	4,102,170	1,324,220	2,777,950	67.72%	1980, 174
1952	4,395,710	1,270,610	3,125,100	71.09%	1980, 174
1953	4,682,420	1,292,890	3,389,530	72.39%	1980, 174
1954	4,803,410	1,393,250	3,410,160	70.99%	1980, 174
1955	4,923,700	1,365,050	3,558,650	72.28%	1980, 174

Appendix Table 3-2. Continued.

Tax Year	Total Filed Individual FIT Returns	Less: Non-taxable FIT Returns	Equals: Taxable FIT Returns[1]	Percent Taxable	Source: RC TS
1956	5,190,751	1,282,575	3,908,176	75.29%	1980, 174
1957	5,478,971	1,402,506	4,076,465	74.40%	1980, 174
1958	5,530,496	1,482,244	4,048,252	73.20%	1980, 174
1959	5,687,525	1,445,035	4,242,490	74.59%	1980, 174
1960	5,850,611	1,460,845	4,389,766	75.03%	1980, 174
1961	5,964,383	1,456,616	4,507,767	75.58%	1980, 174
1962	6,137,227	1,456,000	4,681,227	76.27%	1980, 174
1963	6,350,943	1,423,570	4,927,373	77.58%	1980, 174
1964	6,719,592	1,418,373	5,301,219	78.89%	1980, 174
1965	7,163,160	1,434,218	5,728,942	79.98%	1980, 174
1966	7,733,125	1,456,546	6,276,579	81.16%	1980, 174
1967	8,133,695	1,478,012	6,655,683	81.83%	1980, 174
1968	8,495,184	1,528,270	6,966,914	82.01%	1980, 174
1969	8,882,066	1,518,103	7,363,963	82.91%	1980, 174
1970	9,183,407	1,541,676	7,641,731	83.21%	1980, 174
1971	9,533,292	2,160,721	7,372,571	77.33%	1980, 174
1972	10,382,005	2,300,990	8,081,015	77.84%	1980, 174
1973	11,003,862	2,509,343	8,494,519	77.20%	1980, 174
1974	11,602,170	2,671,938	8,930,232	76.97%	1980, 174
1975	12,002,400	3,510,655	8,491,745	70.75%	1980, 174
1976	12,342,712	3,535,981	8,806,731	71.35%	1980, 174
1977	12,585,891	3,821,173	8,764,718	69.64%	1980, 174
1978	14,320,313	5,507,024	8,813,289	61.54%	1980, 174
1979	14,682,155	5,216,369	9,465,786	64.47%	1992, 69
1980	14,764,878	4,858,036	9,906,842	67.10%	1992, 69
1981	15,179,141	4,748,517	10,430,624	68.72%	1992, 69
1982	15,220,863	4,792,837	10,428,026	68.51%	1992, 69
1983	15,302,940	5,101,540	10,201,400	66.66%	1992, 69
1984	15,552,181	4,901,943	10,650,238	68.48%	1992, 69
1985	15,864,486	4,617,393	11,247,093	70.89%	1992, 69
1986	16,538,060	4,000,440	12,537,620	75.81%	1992, 69
1987	17,071,350	4,007,560	13,063,790	76.52%	1992, 69
1988	17,579,867	4,735,997	12,843,870	73.06%	1992, 69
1989	18,132,050	4,729,870	13,402,180	73.91%	1992, 69
1990	18,758,730	4,962,740	13,795,990	73.54%	1992, 69
1991	19,050,830	5,340,380	13,710,450	71.97%	1997, 273
1992	19,437,070	5,886,270	13,550,800	69.72%	1997, 273
1993	19,829,240	6,260,190	13,569,050	68.43%	1997, 273
1994	20,153,510	6,458,020	13,695,970	67.96%	1997, 273
1995	20,514,590	6,487,920	14,026,670	68.37%	1997, 273
1996	20,805,980	6,633,450	14,172,530	68.12%	1998, 72

Appendix Table 3-2.   Continued.

---

[1] Perry (1955, 698–9) notes that "(p)rior to 1941 tax statistics were compiled for each fiscal year based upon the returns assessed in that fiscal year. These returns were almost entirely those covering the previous taxation year ... for example ... the fiscal year 1928–9 relates to the taxation year 1927 ... Beginning in 1941 the statistics show the actual number of taxpayers in the taxation year".

All measures coincide with those developed by Perry for the 1919–52 tax years, with the exception of 1937 at 264,804, 1938 at 293,097, 1939 at 300,384, 1940 at 387,725, and 1941 at 642,126 (fiscal year-based) or 871,484 (calendar year-based).

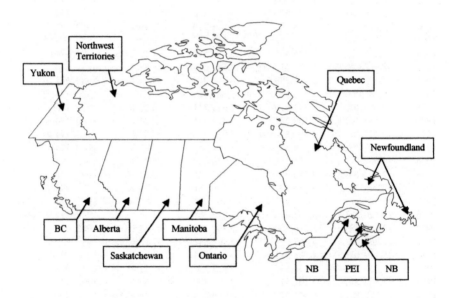

Appendix Fig. 4-1. Canadian Provinces and Territories.

Appendix Table 6-1[1]. Summary of U.S. Personal Exemption (SGL & MFJ), Dependent Allowances, and Standard Deduction (SGL & MFJ) Amounts (1913–2000).

Tax Year	Personal Exemption		Dependent Allowance	Standard Deduction	
	Single	Married		Single	Married

No Dependency Exemption or Standard Deduction (1913–1943):

1913	$3,000	$4,000	$–0–	None	None
1914	$3,000	$4,000	$–0–	None	None
1915	$3,000	$4,000	$–0–	None	None
1916	$3,000	$4,000	$–0–	None	None

Dependency Exemption/Allowance/Deduction Introduced (1917–):

1917	$1,000	$2,000	$200	None	None
1918	$1,000	$2,000	$200	None	None
1919	$1,000	$2,000	$200	None	None
1920	$1,000	$2,000	$200	None	None
1921[2]	$1,000	$2,500	$400	None	None
1922[2]	$1,000	$2,500	$400	None	None
1923[2]	$1,000	$2,500	$400	None	None
1924	$1,000	$2,500	$400	None	None
1925	$1,500	$3,500	$400	None	None
1926	$1,500	$3,500	$400	None	None
1927	$1,500	$3,500	$400	None	None
1928	$1,500	$3,500	$400	None	None
1929	$1,500	$3,500	$400	None	None
1930	$1,500	$3,500	$400	None	None
1931	$1,500	$3,500	$400	None	None
1932	$1,000	$2,500	$400	None	None
1933	$1,000	$2,500	$400	None	None
1934	$1,000	$2,500	$400	None	None
1935	$1,000	$2,500	$400	None	None
1936	$1,000	$2,500	$400	None	None
1937	$1,000	$2,500	$400	None	None
1938	$1,000	$2,500	$400	None	None
1939	$1,000	$2,500	$400	None	None
1940	$800	$2,000	$400	None	None
1941	$750	$1,500	$400	None	None
1942	$500	$1,200	$350	None	None
1943	$500	$1,200	$350	None	None

Variable Standard Deduction at 10% AGI to $1,000 Maximum (1944–1963):

1944[3]	$500	$1,000	$500	10% of AGI from $0-$1,000	
1945	$500	$1,000	$500	10% of AGI from $0-$1,000	

Appendix Table 6-1. Continued.

Tax Year	Personal Exemption		Dependent Allowance	Standard Deduction	
	Single	Married		Single	Married

Variable Standard Deduction at 10% AGI to $1,000 Maximum (1944–1963):

Tax Year	Single	Married	Dependent Allowance	Standard Deduction Single	Married
1946	$500	$1,000	$500	10% of AGI from $0-$1,000	
1947	$500	$1,000	$500	10% of AGI from $0-$1,000	
1948	$600	$1,200	$600	10% of AGI from $0-$1,000	
1949	$600	$1,200	$600	10% of AGI from $0-$1,000	
1950	$600	$1,200	$600	10% of AGI from $0-$1,000	
1951	$600	$1,200	$600	10% of AGI from $0-$1,000	
1952	$600	$1,200	$600	10% of AGI from $0-$1,000	
1953	$600	$1,200	$600	10% of AGI from $0-$1,000	
1954	$600	$1,200	$600	10% of AGI from $0-$1,000	
1955	$600	$1,200	$600	10% of AGI from $0-$1,000	
1956	$600	$1,200	$600	10% of AGI from $0-$1,000	
1957	$600	$1,200	$600	10% of AGI from $0-$1,000	
1958	$600	$1,200	$600	10% of AGI from $0-$1,000	
1959	$600	$1,200	$600	10% of AGI from $0-$1,000	
1960	$600	$1,200	$600	10% of AGI from $0-$1,000	
1961	$600	$1,200	$600	10% of AGI from $0-$1,000	
1962	$600	$1,200	$600	10% of AGI from $0-$1,000	
1963	$600	$1,200	$600	10% of AGI from $0-$1,000	
1964	$600	$1,200	$600	10% of AGI from $200-$1,000	
1965	$600	$1,200	$600	10% of AGI from $200-$1,000	
1966	$600	$1,200	$600	10% of AGI from $200-$1,000	
1967	$600	$1,200	$600	10% of AGI from $200-$1,000	
1968	$600	$1,200	$600	10% of AGI from $200-$1,000	
1969	$600	$1,200	$600	10% of AGI from $200-$1,000	
1970	$625	$1,250	$625	10% of AGI from $200-$1,000	
1971	$675	$1,350	$675	13% of AGI to $1,050	
1972	$750	$1,500	$750	15% of AGI; $1,300-$2,000	
1973	$750	$1,500	$750	15% of AGI; $1,300-$2,000	
1974	$750	$1,500	$750	15% of AGI; $1,300-$2,000	
1975[4]	$750	$1,500	$750	16% of AGI; $1,600-$2,300 (SGL) 16% of AGI; $1,900-$2,600 (MFJ)	
1976	$750	$1,500	$750	16% of AGI; $1,700-$2,400 (SGL) 16% of AGI; $2,100-$2,800 (MFJ)	

Fixed Standard Deduction (1977–):

1977	$750	$1,500	$750	$2,200	$3,200
1978	$750	$1,500	$750	$2,200	$3,200
1979	$750	$1,500	$750	$2,300	$3,400
1980	$1,000	$2,000	$1,000	$2,300	$3,400
1981	$1,000	$2,000	$1,000	$2,300	$3,400

Appendix Table 6-1. Continued.

Tax Year	Personal Exemption		Dependent Allowance	Standard Deduction	
	Single	Married		Single	Married
Fixed Standard Deduction (1977–):					
1982	$1,000	$2,000	$1,000	$2,300	$3,400
1983	$1,000	$2,000	$1,000	$2,300	$3,400
1984	$1,000	$2,000	$1,000	$2,300	$3,400
1985[5]	$1,040	$2,080	$1,040	$2,390	$3,540
1986	$1,080	$2,160	$1,080	$2,480	$3,670
1987	$1,900	$3,800	$1,900	$2,540	$3,760
Inflation-Indexing Reintroduced/Applied to Revised Personal Exemption (1988–):					
1988[6]	$1,950	$3,800	$1,950	$3,000	$5,000
1989	$2,000	$4,000	$2,000	$3,100	$5,200
1990	$2,050	$4,100	$2,050	$3,250	$5,450
Variable Standard Deduction Phase-Out for High-Income Taxpayers (1991–):					
1991[7]	$2,150	$4,300	$2,150	$3,400	$5,700
1992	$2,300	$4,600	$2,300	$3,600	$6,000
1993	$2,350	$4,700	$2,350	$3,700	$6,200
1994	$2,450	$4,900	$2,450	$3,800	$6,350
1995	$2,500	$5,000	$2,500	$3,900	$6,550
1996	$2,550	$5,100	$2,550	$4,000	$6,700
1997	$2,650	$5,300	$2,650	$4,150	$6,900
1998	$2,700	$5,400	$2,700	$4,250	$7,100
1999	$2,750	$5,500	$2,750	$4,300	$7,200
2000	$2,800	$5,600	$2,800	$4,400	$7,350

[1] This table was developed from Brozovsky and Cataldo (1994, 168–9) for 1913–1993. Extensions were made using the IRS SOI for 1994–1996 and the RIA for 1997–2000.

[2] For the 1921–1923 tax years, exemptions declined to $2,000 if *net income* exceeded $5,000.

[3] Beginning with the 1944 tax year, the personal exemption applied to each individual regardless of marital status (e.g. $500 for single and $1,000 for married taxpayers), and dependent allowances were equivalent to the single taxpayer personal exemption.

[4] Beginning with the 1975 tax year, separate standard deduction amounts were available to single and married taxpayers.

[5] First year of inflation-indexing, rounded to $10 increments (1984 base year).

[6] First year of inflation-indexing, rounded to $50 increments (1987 base year).

[7] Beginning with the 1991 tax year, the personal and dependent exemptions for high-income taxpayers were subject to phase-out. These phase-out ranges were presented in Table 1-1.

Appendix Table 6-2[1]. Summary of Canadian Personal Exemptions and Dependent Allowances (1918–1998).

Tax Year	Personal Exemption		Dependent Allowance	Source
	Single	Married		
1917[2]	$1,500	$3,000	None	CYB[3] 1926, 756
1918[4]	$1,000	$2,000	$200	CYB[3] 1926, 756
1919	$1,000	$2,000	$200	CYB[3] 1926, 756
1920	$1,000	$2,000	$200	CYB 1921, 654
1921	$1,000	$2,000	$200	CYB 1921, 654
1922	$1,000	$2,000	$300	CYB 1926, 758
1923	$1,000	$2,000	$500	CYB 1921, 654
1924	$1,000	$2,000	$500	CYB 1926, 758
1925[5]	$1,500	$3,000	$500	CYB 1926, 759
1926[6]	$1,500	$3,000	$500	CYB 1926, 759
1927	$1,500	$3,000	$500	CYB 1927–28, 808
1928	$1,500	$3,000	$500	CYB 1929, 794
1929	$1,500	$3,000	$500	CYB 1931, 833
1930	$1,500	$3,000	$500	CYB 1931, 833
1931	$1,200	$2,400	$500	CYB 1934–35, 824
1932	$1,000	$2,000	$400	CYB 1934–35, 824
1933	$1,000	$2,000	$400	CYB 1934–35, 824
1934	$1,000	$2,000	$400	CYB 1934–35, 824
1935	$1,000	$2,000	$400	CYB 1936, 826
1936	$1,000	$2,000	$400	CYB 1937, 810–11
1937	$1,000	$2,000	$400	CYB 1938, 838–9
1938	$1,000	$2,000	$400	B&K, 372
1939[7]	$1,000	$2,000	$400	B&K, 372
1940	$750	$1,500	$400	DTC[8] 1943, 1
1941	$750	$1,500	$400	TS[9] 1946, 110
1942[10]	$660	$1,200	Tax credit	TS 1949, 108
1943	$660	$1,200	Tax credit	TS 1949, 108
1944	$660	$1,200	Tax credit	TS 1949, 108
1945	$660	$1,200	Tax credit	TS 1949, 108
1946	$660	$1,200	Tax credit	TS 1949, 108
1947[11]	$750	$1,500	$100/$300	TS 1949, 108
1948	$750	$1,500	$100/$300	TS 1950, 112
1949	$1,000	$2,000	$150/$400	TS 1951, 109
1950	$1,000	$2,000	$150/$400	TS 1952, 76
1951	$1,000	$2,000	$150/$400	TS 1953, 28
1952	$1,000	$2,000	$150/$400	TS 1954, 26
1953	$1,000	$2,000	$150/$400	TS 1955, 26
1954	$1,000	$2,000	$150/$400	TS 1956, 26
1955	$1,000	$2,000	$150/$400	TS 1957, 26
1956	$1,000	$2,000	$150/$400	TS 1958, 26

Appendix Table 6-2. Continued.

Tax Year	Personal Exemption		Dependent Allowance	Source
	Single	Married		
1957	$1,000	$2,000	$250/$500	*TS* 1959, 30
1958	$1,000	$2,000	$250/$500	*TS* 1960, 30
1959	$1,000	$2,000	$250/$500	*TS* 1961, 29
1960	$1,000	$2,000	$250/$500	*TS* 1962, 29
1961	$1,000	$2,000	$250/$500	*CYB* 1961, 1039
1962	$1,000	$2,000	$300/$550	*CYB* 1962, 1017
1963	$1,000	$2,000	$300/$550	*CYB* 1963–64, 966
1964	$1,000	$2,000	$300/$550	*CYB* 1963–64, 966
1965	$1,000	$2,000	$300/$550	*CYB* 1965, 958
1966	$1,000	$2,000	$300/$550	*CYB* 1966, 966
1967	$1,000	$2,000	$300/$550	*CYB* 1967, 1021
1968	$1,000	$2,000	$300/$550	*CYB* 1968, 1011
1969	$1,000	$2,000	$300/$550	*CYB* 1969, 1040
1970	$1,000	$2,000	$300/$550	*CYB* 1970, 1127
1971[12]	$1,000	$2,000	$300/$550	*CTF*[13] 1986–87, 7:6
1972[14]	$1,500	$2,850	$300/$550	*CTF* 1986–87, 7:6
1973	$1,600	$3,000	$300/$550	*CTF* 1986–87, 7:6
1974[15]	$1,706	$3,198	$320/$586	*CTF* 1986–87, 7:6
1975	$1,878	$3,522	$352/$646	*CTF* 1986–87, 7:6
1976	$2,090	$3,920	$392/$719	*CTF* 1986–87, 7:6
1977	$2,270	$4,260	$430/$780	*CTF* 1986–87, 7:6
1978	$2,430	$4,560	$460/$840	*CTF* 1986–87, 7:6
1979	$2,650	$4,970	$500/$910	*CTF* 1986–87, 7:6
1980[16]	$2,890	$5,420	$540/$990	*CTF* 1986–87, 7:6
1981	$3,170	$5,950	$590/$1,090	*CTF* 1986–87, 7:6
1982	$3,560	$6,670	$670/$1,220	*CTF* 1986–87, 7:6
1983	$3,770	$7,070	$710/$1,300	*CTF* 1986–87, 7:6
1984[17]	$3,960	$7,430	$710/$1,360	*CTF* 1986–87, 7:6
1985	$4,140	$7,770	$710/$1,420	*CTF* 1986–87, 7:6
1986[18]	$4,180	$7,840	$710/$1,420	*CTF* 1986–87, 7:6
1987	$4,220	$7,920	$560/$1,200	*CTF* 1986–87, 7:6
1988[19]	$4,270	$8,010	$388/$1,000	*CTF* 1987–88, 7:8
1989	$6,066	$11,120	$392/$784	*CTF* 1990, 7:6
1990	$6,169	$11,310	$399/$798	*CTF* 1990, 7:6
1991	$6,280	$11,513	$406/$812	*CTF* 1991, 7:5
1992	$6,456	$11,836	$417/$834	*CTF* 1992, 7:9
1993[20]	$6,456	$11,836	None	*CTF* 1993, 7:8–9
1994	$6,456	$11,836	None	*CTF* 1995, 3:8
1995	$6,456	$11,836	None	*CTF* 1995, 3:8
1996	$6,456	$11,836	None	1996 RC Tax Forms
1997	$6,456	$11,836	None	1997 RC Tax Forms
1998	$6,456	$11,836	None	1998 RC Tax Forms

Appendix Table 6-2.   Continued.

---

[1] Separate reference sources are cited and were verified, for the 1917 through 1983 tax years, using Boadway and Kitchen (B&K; 1984, 372–5).

[2] The Canadian Parliament declared war (WWI) on Germany (August 4, 1914), Austria-Hungary (August 12, 1914), and Turkey (November 5, 1914).

[3] *Canada Year Book (CYB).*

[4] Beginning with the 1918 tax year, a dependent exemption was granted for each (taxpayer's own) child under 18 years of age and dependent on the taxpayer for support.

[5] Beginning with the 1925 tax year, the dependent exemption was expanded to include non-familial dependents.

[6] Beginning with the 1926 tax year, the dependent exemption was expanded to include those under 21 years of age.

[7] The Canadian Parliament declared war (WWII) on Germany on September 10, 1939.

[8] *Department of Trade and Commerce (DTC).*

[9] *Taxation Statistics (TS).*

[10] The form of the dependent allowance was changed from a tax *deduction* to a tax *credit* for the 1942 through 1946 tax years, for dependents under 18 or under 21 and attending school. The maximum tax credit was $108 ($28 for the *normal* tax and $80 for the *graduated* or *progressive* tax components). Therefore, the personal exemption was effectively reduced from $750 in 1941 to $660 in 1942 for persons taxed as single and from $1,500 in 1941 to $1,200 in 1942 for persons taxed as married (i.e. a tax credit of $150 was applied to persons having married status, resulting in an effective personal exemption amount of $1,200 for persons taxed as married). Also, beginning with the 1942 tax year, an additional exemption became available for taxpayers over 65 years of age.

[11] Beginning with the 1947 tax year, the form of the dependent allowance was restored from a tax *credit* to a tax *deduction*. The dependent allowance, beginning with the 1947 tax year, differed for those eligible for family allowance payments ($100 for 1947) and those not eligible ($300 for 1947).

[12] Beginning with the 1971 tax year, an additional exemption was provided for blind or incapacitated taxpayers.

[13] *Canadian Tax Foundation (CTF).*

[14] Beginning with the 1972 tax year, an educational exemption was provided. Also beginning with the 1972 tax year, different *single status* and *married additional* personal exemptions were provided. Finally, for the 1972 through 1978 tax years, the dependent exemption was expanded to include dependents under 16 years of age. Dependent allowances differed for dependents under the age of 16 and dependents 16 and over.

[15] Beginning with the 1974 tax year, inflation-indexing was introduced for both personal exemptions and dependent allowances.

[16] The 18 year threshold was re-established, beginning with the 1980 tax year.

[17] Beginning with the 1984 tax year, inflation-indexing was suspended for the dependency exemption for children under 18 years of age.

[18] Beginning with the 1986 tax year, inflation-indexing was suspended for the dependency exemption for children 18 years of age or older.

Appendix Table 6-2.   Continued.

[19] Beginning with the 1988 tax year, personal exemption tax *credits* replaced tax *deductions*. However, for consistency, this table continues to state the personal exemption in the form of a deduction. Also new for the 1988 tax year, dependent exemption allowances differed for the first two dependent children under the age of 19 and *other* dependents.

[20] Beginning with the 1993 tax year, the *dependent allowance* (included in this table), as well as the *child credit* and the *refundable child tax credit* (not presented in this table) were replaced with the inflation-indexed (for annual inflation rates in excess of 3% only) *child tax benefit*. To illustrate the conversion of deductions to credits (in this table): the personal exemption for the single or *individual* taxpayer (at $6,456) represents a tax credit of $1,098 ($6,456 *multiplied by* 17% – the lowest marginal tax bracket). Similarly, the personal exemption for the *spouse* (at $5,380) represents a tax credit of $915 ($5,380 *multiplied by* 17% – the lowest marginal tax bracket). The total is $11,836 ($6,456 *plus* $5,380).

Appendix Table 7-1[1].   The Benefits Structure of the U.S. Post-1974 Earned
Income Tax Credit (EITC).

Tax Year	Number of Qualify Dependents	Credit Rate	Phase-out Rate	"Flat" Range Begins	"Flat" Range Ends	Max EI or AGI	Max EITC	Source
Phase I: Initial Implementation:								
1975	≥ 1	10%	10%	$4,000	$4,000	$8,000	$400	Cataldo 1995, 69
1976	≥ 1	10%	10%	$4,000	$4,000	$8,000	$400	Cataldo 1995, 69
1977	≥ 1	10%	10%	$4,000	$4,000	$8,000	$400	Cataldo 1995, 69
1978	≥ 1	10%	10%	$4,000	$4,000	$8,000	$400	Cataldo 1995, 69
1979	≥ 1	10%	12.5%	$5,000	$6,000	$10,000	$500	Cataldo 1995, 69
1980	≥ 1	10%	12.5%	$5,000	$6,000	$10,000	$500	Cataldo 1995, 69
1981	≥ 1	10%	12.5%	$5,000	$6,000	$10,000	$500	Cataldo 1995, 69
1982	≥ 1	10%	12.5%	$5,000	$6,000	$10,000	$500	Cataldo 1995, 69
1983	≥ 1	10%	12.5%	$5,000	$6,000	$10,000	$500	Cataldo 1995, 69
1984	≥ 1	10%	12.5%	$5,000	$6,000	$10,000	$500	Cataldo 1995, 69
1985	≥ 1	11%	12.22%	$5,000	$6,500	$11,000	$550	Cataldo 1995, 69
1986	≥ 1	11%	12.22%	$5,000	$6,500	$11,000	$550	Cataldo 1995, 69
1987	≥ 1	14%	10%	$6,080	$6,920	$15,432	$851	Cataldo 1995, 69
1988	≥ 1	14%	10%	$6,240	$9,840	$18,576	$874	Cataldo 1995, 69
1989	≥ 1	14%	10%	$6,500	$10,240	$19,340	$910	Cataldo 1995, 69
1990	≥ 1	14%	10%	$6,810	$10,734	$20,264	$953	Cataldo 1995, 69
Phase II: Supplemental Health Care and Newborn Components Added:								
1991	1	16.7%	11.93%	$7,140	$11,250	$21,250	$1,192	Cataldo 1995, 69
	2	17.3%	12.36%	$7,140	$11,250	$21,250	$1,235	Cataldo 1995, 69
	Health	6.0%	4.285%	$7,140	$11,250	$21,250	$428	Cataldo 1995, 69
	Newborn	5.0%	3.57%	$7,140	$11,250	$21,250	$357	Cataldo 1995, 69
1992	1	17.6%	12.57%	$7,520	$11,840	$22,370	$1,324	Cataldo 1995, 69
	2	18.4%	13.14%	$7,520	$11,840	$22,370	$1,384	Cataldo 1995, 69
	Health	6.0%	4.285%	$7,520	$11,840	$22,370	$451	Cataldo 1995, 69
	Newborn	5.0%	3.57%	$7,520	$11,840	$22,370	$376	Cataldo 1995, 69
1993	1	18.5%	13.21%	$7,750	$12,200	$23,050	$1,434	Cataldo 1995, 69
	2	19.5%	13.93%	$7,750	$12,200	$23,050	$1,511	Cataldo 1995, 69
	Health	6.0%	4.285%	$7,750	$12,200	$23,050	$465	Cataldo 1995, 69
	Newborn	5.0%	3.57%	$7,750	$12,200	$23,050	$388	Cataldo 1995, 69
Phase III: Supplemental Components Combined with the Basic Credit & Inclusion of Low-Income Taxpayers without Dependents:								
1994	0	7.65%	7.65%	$4,000	$5,000	$9,000	$306	Cataldo 1995, 69
	1	26.30%	15.98%	$7,750	$11,000	$23,753	$2,038	Cataldo 1995, 69
	2	30.00%	17.68%	$8,425	$11,000	$25,300	$2,528	Cataldo 1995, 69

Appendix Table 7-1. Continued.

Tax Year	Number of Qualify Dependents	Credit Rate	Phase-out Rate	"Flat" Range Begins	"Flat" Range Ends	Max EI or AGI	Max EITC	Source
Phase III: Supplemental Components Combined with the Basic Credit & Inclusion of Low-Income Taxpayers without Dependents:								
1995	0	7.65%	7.65%	$4,168	$5,210	$9,378	$319	Cataldo 1995, 69
	1	34.00%	15.98%	$6,252	$11,462	$24,764	$2,126	Cataldo 1995, 69
	2	36.00%	20.22%	$8,779	$11,462	$27,090	$3,160	Cataldo 1995, 69
1996	0	7.65%	7.65%	$4,343	$5,429	$9,772	$332	Cataldo 1995, 69
	1	34.00%	15.98%	$6,515	$11,943	$25,804	$2,215	Cataldo 1995, 69
	2	40.00%	21.06%	$9,148	$11,943	$29,318	$3,659	Cataldo 1995, 69
1997	0	7.65%	7.65%	$4,340	$5,430	$9,770	$332	*RIA* 1997, 388
	1	34.00%	15.98%	$6,500	$11,930	$25,760	$2,210	*RIA* 1997, 388
	2	40.00%	21.06%	$9,140	$11,930	$29,290	$3,656	*RIA* 1997, 388
1998	0	7.65%	7.65%	$4,460	$5,570	$10,030	$341	*RIA* 1998, 406
	1	34.00%	15.98%	$6,680	$12,260	$26,473	$2,271	*RIA* 1998, 406
	2	40.00%	21.06%	$9,390	$12,260	$30,095	$3,756	*RIA* 1998, 406
1999	0	7.65%	7.65%	$4,530	$5,670	$10,200	$347	*RIA* 1999, 406–7
	1	34.00%	15.98%	$6,800	$12,460	$26,928	$2,312	*RIA* 1999, 406–7
	2	40.00%	21.06%	$9,540	$12,460	$30,580	$3,816	*RIA* 1999, 406–7

[1] This framework is adapted from Cataldo (1995).

Appendix Table 7-2. Summary Statistics of Three Variations of the U.S. Earned Income Credit (EIC) (1924–1931, 1934–1943, and 1975–).

Tax Year	Tax Returns Affected	Percent Affected	Gross EIC	Reference: *IRS SOI*
The FIRST Earned Income Credit (1923–1931):				
Applied Against Tax – Similar to the Contemporary Tax Credit				
1923[1]	Unknown	Unknown	$220,555,000	1931, 38
Applied to all net income ≤ $5,000 & ≤ $10,000 if earned:				1943, 358
1924	Unknown	Unknown	$30,637,463	1924, 7 & 106
Applied to all net income ≤ $5,000 & ≤ $20,000 if earned:				1943, 358
1925	Unknown	Unknown	$24,570,183	1925, 5 & 89
1926	Unknown	Unknown	$24,646,993	1926, 6 & 76
1927	Unknown	Unknown	$24,915,315	1927, 6–7 & 69
Applied to all net income ≤ $5,000 & ≤ $30,000 if earned:				1943, 358
1928	Unknown	Unknown	$34,789,690	1928, 7 & 75
1929	Unknown	Unknown	$22,062,492	1929, 8 & 62
1930	Unknown	Unknown	$24,886,344	1930, 8–9 & 70
1931	Unknown	Unknown	$17,490,530	1931, 8–9 & 61
The SECOND Earned Income Credit (1934–1943)				
Applied Against Net (Taxable) Income – Unlike a Contemporary Tax Credit[2]				
Applied to all net income ≤ $3,000 & ≤ $14,000 if earned:				1943, 358
1934	Unknown	Unknown	$875,962,286	1934, 61
1935	Unknown	Unknown	$944,357,524	1935, 75
1936	Unknown	Unknown	$1.231B	1936, 85
1937	Unknown	Unknown	$1.481B	1937, 118
1938	Unknown	Unknown	$1.391B	1938, 11
1939	Unknown	Unknown	$1.743B	1939, 9
1940	Unknown	Unknown	$3.029B	1940, 91
1941	Unknown	Unknown	$4.736B	1941, 77
1942	Unknown	Unknown	Not Available	
1943	Unknown	Unknown	Not Available	
The THIRD (and CONTEMPORARY) Earned Income Credit (1975–):				
1975	6,214,533	7.56%	$2.977B	1975, 69
1976	6,472,633	7.64%	$3.018B	1976, 105
1977	5,626,938	6.50%	$2.613B	1977, 118
1978	5,192,834	5.78%	$2.436B	1978, 95
1979	7,134,756	7.70%	$4.204B	1979, 80
1980	6,953,621	7.41%	$4.024B	1980, 88
1981	6,717,177	7.04%	$3.873B	1981, 85

Appendix Table 7-2. Continued.

Tax Year	Tax Returns Affected	Percent Affected	Gross EIC	Reference: *IRS SOI*
The <u>THIRD</u> (and <u>CONTEMPORARY</u>) Earned Income Credit (1975–):				
1982	6,395,032	6.71%	$1.775B	1982, 82
1983	Unknown	Unknown	Not Available[3]	
1984	7,109,451	7.15%	$1.636B	1984, 32–5
1985	8,206,704	8.07%	$2.088B	1985, 33–5
1986	7,823,113	7.38%	$2.007B	1986, 36, 39 & 40
1987	10,183,563	9.52%	$3.571B	1987, 58 & 62
1988	11,148,476	10.16%	$5.896B	1992, 2 & 4
1989	11,695,876	10.43%	$6.576B	1993, 2 & 4
1990	12,554,681	11.04%	$7.512B	1993, 2 & 4
1991	13,664,555	11.91%	$11.105B	1993, 2 & 4
1992	14,096,575	12.41%	$13.028B	1993, 2 & 4
1993	15,117,389	13.19%	$15.537B	1993, 2 & 4
1994	19,017,357	16.40%	$21.105B	1994, 86
1995	19,334,397	16.35%	$25.956B	1995, 67
1996	19,463,836	16.17%	$28.825B	1996, 67

[1] For the 1923 tax year, a 25% reduction in tax was provided for in IRC Section 1200 (a) of the Revenue Act of 1924.

[2] This earned income tax "credit" evolved into a *variable* standard deduction (1944–1963), a *semi-variable* standard deduction (1964–1976), and the contemporary *fixed*, but inflation-indexed, standard deduction (1977–).

[3] Only net measures reported for the 1983 tax year.

Appendix Table 8-1. Maximum U.S. Marriage Tax Penalty (MTP) by 1st Quartile, Weighted Average, and 3rd Quartile AGI – in Nominal[1] and 1998 CPI-Adjusted Dollars (1914–1996).

Tax Year	Nominal Marriage Tax Penalty Measures			CPI-Adjusted Marriage Tax Penalty		
	Q1	Weighted Average	Q3	Q1	Weighted Average	Q3
1914	$5	$20	$20	$82	$330	$330
1915	$5	$20	$20	$82	$327	$327
1916	$10	$40	$40	$151	$605	$605
1917	$–0–	$3	$–0–	$–0–	$39	$–0–
1918-1928	*No Incidence of Marriage Tax Penalty*					
1929	$9	$–0–	$–0–	$86	$–0–	$–0–
1930–1946	*No Incidence of Marriage Tax Penalty*					
1947	$–0–	$–0–	$3	$–0–	$–0–	$22
1948-1963	*No Incidence of Marriage Tax Penalty*					
1964	$24	$–0–	$–0–	$126	$–0–	$–0–
1965	$22	$–0–	$–0–	$114	$–0–	$–0–
1966	$22	$–0–	$–0–	$111	$–0–	$–0–
1967	$22	$–0–	$–0–	$107	$–0–	$–0–
1968	$24	$–0–	$59	$112	$–0–	$276
1969	$24	$–0–	$61	$107	$–0–	$271
1970	$8	$–0–	$34	$34	$–0–	$143
1971	$148	$171	$143	$595	$688	$575
1972	$98	$213	$164	$382	$830	$639
1973	$98	$194	$146	$360	$712	$536
1974	$98	$175	$156	$324	$579	$516
1975	$95	$261	$198	$288	$791	$600
1976	$54	$271	$274	$155	$776	$785
1977	$–0–	$244	$376	$–0–	$656	$1,012
1978	$42	$280	$376	$105	$700	$940
1979	$14	$275	$478	$31	$618	$1,074
1980	$14	$304	$478	$28	$602	$946
1981	$154	$365	$747	$276	$655	$1,340
1982	$113	$258	$471	$191	$436	$796
1983	$85	$142	$292	$139	$232	$478
1984	$91	$115	$253	$143	$181	$397
1985	$95	$123	$443	$144	$186	$671
1986	$101	$133	$448	$150	$198	$666
1987	$–0–	$222	$222	$–0–	$319	$319

Appendix Table 8-1. Continued.

Tax Year	Nominal Marriage Tax Penalty Measures			CPI-Adjusted Marriage Tax Penalty		
	Q1	Weighted Average	Q3	Q1	Weighted Average	Q3
1988	$–0–	$150	$150	$–0–	$207	$207
1989	$–0–	$150	$150	$–0–	$197	$197
1990	$–0–	$161	$153	$–0–	$201	$191
1991	$–0–	$164	$162	$–0–	$196	$194
1992	$–0–	$184	$176	$–0–	$214	$205
1993	$–0–	$184	$177	$–0–	$208	$200
1994	$–0–	$192	$184	$–0–	$211	$202
1995	$–0–	$192	$183	$–0–	$206	$196
1996	$–0–	$199	$192	$–0–	$207	$200

[1] As adapted from Brozovsky and Cataldo (1994, 186–187). Estimated measures from 1990–1993 were replaced and extended through 1996.

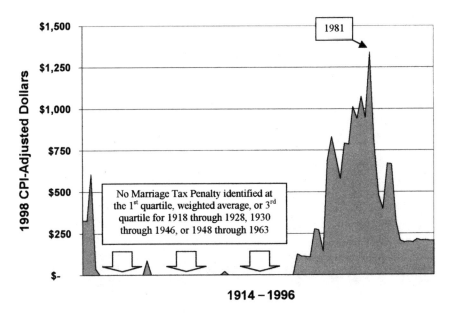

Appendix Fig. 8-1. Maximum U.S. Marriage Tax Penalty.

Appendix Table 8-2.   Maximum U.S. Marriage Tax Bonus (MTB) by 1st Quartile, Weighted Average, and 3rd Quartile AGI – in Nominal[1] and 1998 CPI-Adjusted Dollars (1914–1996).

Tax Year	Nominal Marriage Tax Bonus Measures			CPI-Adjusted Marriage Tax Bonus		
	Q1	Weighted Average	Q3	Q1	Weighted Average	Q3
1914	$10	$10	$10	$165	$165	$165
1915	$10	$10	$10	$163	$163	$163
1916	$20	$20	$20	$303	$303	$303
1917	$10	$40	$40	$129	$516	$516
1918	$30	$60	$60	$328	$655	$655
1919	$20	$40	$40	$191	$381	$381
1920	$20	$40	$40	$165	$330	$330
1921	$20	$60	$60	$184	$553	$553
1922	$20	$60	$60	$196	$589	$589
1923	$11	$60	$34	$106	$578	$328
1924	$8	$23	$23	$77	$222	$222
1925	$–0–	$23	$23	$–0–	$214	$214
1926	$–0–	$24	$23	$–0–	$221	$212
1927	$–0–	$26	$23	$–0–	$244	$216
1928	$–0–	$34	$23	$–0–	$319	$216
1929	$9	$33	$45	$86	$315	$429
1930	$–0–	$23	$23	$–0–	$224	$224
1931	$–0–	$23	$23	$–0–	$247	$247
1932	$20	$60	$60	$238	$714	$714
1933	$20	$60	$60	$251	$752	$752
1934	$14	$60	$60	$170	$730	$730
1935	$14	$60	$60	$167	$714	$714
1936	$14	$60	$60	$164	$703	$703
1937	$14	$60	$60	$158	$679	$679
1938	$50	$60	$50	$578	$694	$578
1939	$50	$60	$50	$586	$703	$586
1940	$24	$53	$53	$279	$617	$617
1941	$69	$75	$75	$765	$832	$832
1942	$133	$139	$133	$1,330	$1,390	$1,330
1943	$158	$187	$198	$1,489	$1,762	$1,866
1944	$100	$100	$113	$927	$927	$1,047
1945	$100	$100	$100	$906	$906	$906
1946	$95	$95	$95	$795	$795	$795
1947	$95	$96	$105	$695	$702	$768
1948	$87	$110	$115	$588	$744	$778
1949	$87	$109	$115	$596	$746	$787
1950	$104	$116	$136	$703	$784	$919

Appendix Table 8-2.   Continued.

Tax Year	Nominal Marriage Tax Bonus Measures			CPI-Adjusted Marriage Tax Bonus		
	Q1	Weighted Average	Q3	Q1	Weighted Average	Q3
1951	$122	$137	$147	$764	$858	$921
1952	$133	$154	$173	$817	$946	$1,063
1953	$133	$158	$205	$811	$963	$1,250
1954	$120	$142	$181	$727	$860	$1,096
1955	$120	$146	$181	$729	$888	$1,100
1956	$120	$151	$181	$719	$904	$1,084
1957	$120	$154	$181	$696	$893	$1,049
1958	$120	$156	$222	$677	$880	$1,252
1959	$120	$164	$222	$672	$918	$1,243
1960	$120	$170	$222	$661	$936	$1,223
1961	$120	$178	$222	$654	$971	$1,210
1962	$120	$189	$264	$648	$1,020	$1,425
1963	$120	$199	$264	$640	$1,061	$1,407
1964	$109	$220	$324	$573	$1,157	$1,705
1965	$103	$227	$307	$533	$1,175	$1,590
1966	$103	$235	$307	$518	$1,183	$1,545
1967	$103	$248	$361	$503	$1,210	$1,762
1968	$111	$285	$636	$520	$1,335	$2,979
1969	$113	$318	$651	$502	$1,412	$2,890
1970	$138	$328	$513	$580	$1,378	$2,155
1971	$117	$306	$435	$471	$1,231	$1,750
1972	$119	$333	$431	$464	$1,298	$1,680
1973	$119	$364	$473	$437	$1,336	$1,736
1974	$119	$398	$523	$394	$1,316	$1,730
1975	$214	$443	$770	$649	$1,343	$2,334
1976	$222	$449	$757	$636	$1,286	$2,169
1977	$199	$702	$921	$535	$1,889	$2,478
1978	$322	$767	$921	$805	$1,918	$2,303
1979	$318	$776	$1,330	$714	$1,743	$2,988
1980	$318	$907	$1,330	$629	$1,795	$2,632
1981	$358	$1,009	$1,744	$642	$1,811	$3,129
1982	$316	$992	$1,589	$534	$1,676	$2,685
1983	$295	$936	$1,419	$483	$1,532	$2,323
1984	$304	$1,020	$1,325	$477	$1,601	$2,080
1985	$308	$1,077	$1,804	$467	$1,632	$2,733
1986	$311	$1,137	$1,793	$462	$1,691	$2,666
1987	$387	$1,226	$2,554	$555	$1,760	$3,666
1988	$533	$1,329	$2,179	$735	$1,832	$3,004
1989	$510	$1,396	$2,091	$671	$1,837	$2,751

Appendix Table 8-2.   Continued.

Tax Year	Nominal Marriage Tax Bonus Measures			CPI-Adjusted Marriage Tax Bonus		
	Q1	Weighted Average	Q3	Q1	Weighted Average	Q3
1990	$484	$1,408	$1,974	$604	$1,757	$2,464
1991	$444	$1,399	$1,854	$532	$1,676	$2,221
1992	$544	$1,404	$3,003	$633	$1,633	$3,493
1993	$521	$1,413	$2,921	$588	$1,595	$3,298
1994	$491	$1,488	$2,834	$540	$1,638	$3,119
1995	$469	$1,634	$2,759	$502	$1,749	$2,954
1996	$596	$1,785	$2,669	$619	$1,855	$2,774

[1] As adapted from Brozovsky and Cataldo (1994, 186–187). Estimated measures from 1990–1993 were replaced and extended through 1996.

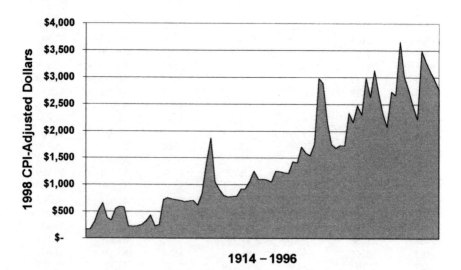

Fig. 8-2. Maximum U.S. Marriage Tax Bonus.

Appendix Table 8-3.   U.S. Weighted Average AGI – in Nominal[1] and 1998
CPI-Adjusted Dollars (1914–1996).

Tax Year	Tax Nominal	Adjusted	Year	Nominal	Adjusted
1914	$10,826	$178,541	1956	$4,612	$27,622
1915	$12,993	$212,157	1957	$4,786	$27,748
1916	$14,733	$222,956	1958	$4,865	$27,438
1917	$4,150	$53,494	1959	$5,189	$29,062
1918	$3,840	$41,948	1960	$5,293	$29,148
1919	$3,983	$37,967	1961	$5,451	$29,721
1920	$3,483	$28,720	1962	$5,653	$30,518
1921	$3,134	$28,874	1963	$5,870	$31,282
1922	$3,343	$32,801	1964	$6,385	$33,590
1923	$3,419	$32,953	1965	$6,699	$34,687
1924	$3,677	$35,440	1966	$6,995	$35,199
1925	$5,526	$51,460	1967	$7,405	$36,142
1926	$5,598	$51,563	1968	$7,729	$36,203
1927	$5,813	$54,470	1969	$8,176	$36,300
1928	$6,499	$60,898	1970	$8,685	$36,480
1929	$6,406	$61,064	1971	$9,234	$37,152
1930	$5,197	$50,706	1972	$9,845	$38,382
1931	$4,496	$48,205	1973	$10,528	$38,648
1932	$3,211	$38,210	1974	$11,192	$37,014
1933	$3,156	$39,574	1975	$11,912	$36,110
1934	$3,351	$40,756	1976	$12,893	$36,941
1935	$3,495	$41,592	1977	$13,892	$37,374
1936	$3,820	$44,788	1978	$15,127	$37,822
1937	$3,590	$40,629	1979	$16,178	$36,343
1938	$3,293	$38,067	1980	$17,628	$34,890
1939	$3,276	$38,408	1981	$19,145	$34,354
1940	$2,603	$30,306	1982	$20,304	$33,854
1941	$2,747	$30,459	1983	$20,806	$34,065
1942	$2,712	$27,115	1984	$22,210	$34,864
1943	$3,056	$28,798	1985	$23,481	$35,579
1944	$2,551	$23,638	1986	$24,838	$36,933
1945	$2,479	$22,454	1987	$26,703	$38,327
1946	$2,622	$21,929	1988	$28,466	$39,248
1947	$2,817	$20,594	1989	$29,656	$39,016
1948	$3,293	$22,270	1990	$30,696	$38,315
1949	$3,254	$22,274	1991	$31,535	$37,776
1950	$3,402	$22,988	1992	$32,718	$38,051
1951	$3,676	$23,021	1993	$33,420	$37,736
1952	$3,867	$23,766	1994	$34,687	$38,174
1953	$4,022	$24,522	1995	$36,395	$38,963
1954	$4,107	$24,866	1996	$38,240	$39,745
1955	$4,334	$26,346			

[1] As adapted from Brozovsky and Cataldo (1994, 186–187). Estimated measures from 1990–1993 were replaced and extended through 1996.

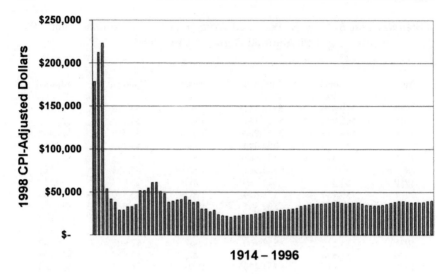

Appendix Fig. 8-3. U.S. Weighted Average AGI.

Appendix Table 8-4. Summary of Selected Contemporary Studies of the Marriage Tax.

Author(s)	Methodology	Scope	Findings
Descriptive Studies:			
Rosen (1987)	Divorce-based estimates of MTP	1983 U.S. data projected to 1988	Calculation of MTP
Feenberg & Rosen (1995)	Divorce-based estimates of MTP	1989 U.S. data projected to 1994	Calculation of MTP
Cataldo (1996)	Decomposition of MTP components	1989 U.S. data	Decomposition of MTP components
Predictive Studies:			
Sjoquist & Walker (1995)	Marriage-based	U.S. 1948-87 data	MTP affects only *timing* of marriage
Alm & Whittington (1995)	Actual divorce & marriage rates	U.S. 1947–88 data	MTP affects both *timing & incidence* of marriage
Gelardi (1996)	Actual divorce & marriage rates	Canada 1950–91 & England/Wales 1960–91 data	MTP affects *timing* of marriage; tests not designed to detect *incidence* of marriage

Appendix Table 9-1. The First U.S. *Alternative* to Ordinary Tax Rates (1922–) for LTCG and *Credits*[1] (1924–) for LTCL: The $12\frac{1}{2}$ Percent Alternative Tax AT (000 omitted).

1922–1933

Tax Year	Exempt from *Normal* Tax	Tax Before *Credits* at $12\frac{1}{2}$ Percent	Tax *Credit* for LTCL	Source: *IRS SOI*
1922	$249,248	$31,066	n.a.	1934, 23 & 27–8
1923	$305,394	$38,916	n.a.	1934, 23 & 27–8

Tax Credit of $12\frac{1}{2}$ Percent Provided for LTCL:

1924	$389,148	$48,603	$9,036	1934, 23 & 27–8
1925	$940,569	$117,571	$7,659	1934, 23 & 27–8
1926	$912,917	$112,510	$4,322	1934, 23 & 27–8
1927	$1,081,186	$134,034	$6,028	1934, 23 & 27–8
1928	$1,879,780	$233,451	$5,126	1934, 23 & 27–8
1929	$2,346,704	$284,654	$5,378	1934, 23 & 27–8

Post–1929 Stock Market "Crash" Tax Years:

1930	$556,392	$65,422	$10,112	1934, 23 & 27–8
1931	$169,949	$19,423	$24,185	1934, 23 & 27–8
1932	$50,074	$6,039	$71,915	1934, 23 & 27–8
1933	$133,616	$16,435	$50,899	1934, 23 & 27–8

[1] During this historical period, the term *credit* represented what we presently refer to as a *deduction*.

Appendix Table 9-2. The Second U.S. Alternative Tax (AT) and the Optional Tax, also Designated as an AT.

Tax Year	Tax Returns Affected	Percent Affected	Source: *IRS SOI*
1938	6,796	00.11%	1938, 86
1939	8,365	00.11%	1939, 86
1940	25,286	00.17%	1940, 97
1941	68,176	00.26%	1941, 97
1942	12,946	00.04%	1942, 93
1943	31,850	00.07%	1943, 18
1944	51,993	00.11%	1944, 100
1945	88,485	00.18%	1945, 105
1946	84,021	00.16%	1946, 108
1947	69,444	00.13%	1947, 174
1948	30,896	00.06%	1948, 29
1949	25,139	00.05%	1949, 30
1950	49,316	00.09%	1950, 116
1951	70,655	00.13%	1951, 56
1952	80,700	00.14%	1952, 33
1953	68,665	00.12%	1953, 41
1954	73,618	00.13%	1954, 56–7
1955	91,014	00.16%	1955. 35–7
1956	86,499	00.15%	1956, 42–3
1957	76,413	00.13%	1957, 37
1958	88,941	00.15%	1958, 50–1
1959	110,296	00.18%	1959, 44–6
1960	91,818	00.15%	1960, 63–4
1961	108,759	00.18%	1961, 74
1962	89,484	00.14%	1962, 12
1963	143,524	00.22%	1963, 59
1964	99,679	00.15%	1964, 32
1965	86,804	00.13%	1965, 29
1966	85,671	00.12%	1966, 94
1967	114,606	00.16%	1967, 75
1968	136,841	00.19%	1968, 120
1969	101,260	00.15%	1969, 120
1970	69,705	00.09%	1970, 157
1971	77,319	00.10%	1971, 138
1972	95,894	00.12%	1972, 154
1973	90,854	00.11%	1973, 79
1974	76,317	00.09%	1974, 126
1975	90,688	00.11%	1975, 74
1976	142,098	00.17%	1976, 81
1977	355,268	00.41%	1977, 82
1978	Not available	Not available	

Appendix Table 9-3.   The Minimum Tax (MINTAX) on "Tax Preferences".

Tax Year	Tax Returns Affected	Percent Affected	Source: *IRS SOI*
TRA69 Introduced a 10% MINTAX on "Tax Preferences" (1970–):			
1970	18,942	00.03%	1970, 150
1971	23,606	00.03%	1971, 135
1972	26,618	00.03%	1972, 180
1973	26,382	00.03%	1973, 119
1974	18,542	00.02%	1974, 120
1975	20,188	00.02%	1975, 91
TRA76 Increased the MINTAX from 10% to 15% (1976–):			
1976	246,735	00.29%	1976, 108
1977	399,487	00.46%	1977, 73
1978	495,332	00.55%	1978, 100
TRA78 Established the Contemporary Alternative Minimum Tax (AMT) (1979–); Taxpayers Required to Pay the GREATER of the AMT or Taxes Otherwise Computed:			
1979	74,421	00.08%	1979, 61
1980	94,599	00.10%	1980, 7
1981	125,721	00.13%	1981, 7
1982	101,423	00.11%	1982, 9
The AMT is Raised to a Rate of 20%:			
1983[1]	1,419	00.00%	1983, 25

[1] These measures relate to 1982 tax returns.

Appendix Table 9-4.   The Contemporary U.S. Alternative Minimum Tax (AMT).

Tax Year	Tax Returns Affected	Percent Affected	AMT Tax Generated	Source: *IRS SOI*
1979	153,265[1]	00.17%	$0.865940B	1979, 59 & 87
1980	122,670	00.13%	$0.850326B	1979, 59 & 87
1981	137,113	00.14%	$1.261318B	1979, 59 & 87
1982	131,376	00.14%	$1.069214B	1979, 59 & 87
1983	264,690	00.27%	$2.520954B	1979, 59 & 87
1984	370,212	00.37%	$4.490251B	1984, 33
1985	427,688	00.42%	$3.791672B	1985, 34
1986	608,907	00.59%	$6.713149B	1986, 38
1987	139,779	00.13%	$1.674898B	1987, 37
1988	113,562	00.10%	$1.027884B	1988, 36
1989	117,483	00.10%	$0.830994B[2]	1993, 51
1990	132,098[2]	00.12%	$0.830311B[2]	1993, 51
1991	243,672	00.21%	$1.213426B	1993, 58
1992	287,183	00.25%	$1.357063B	1993, 91
1993	334,615	00.29%	$2.052790B	1993, 4
1994	368,964	00.32%	$2.212094B	1996, 4
1995	414,106	00.35%	$2.290576B	1996, 4
1996	477,898	00.40%	$2.812746B	1996, 4

[1] Approximately 133,246 returns (87%) of these individual tax returns with AMT were subject *only* to the AMT, for a total AMT of $648.682 million (75%).

[2] The IRS SOI (1993, 4) provides a comparable measure at 132,103 returns for the 1990 tax year. Similarly, the tax revenue measures for the 1989 and 1990 tax years are comparable (i.e. $831.012 million for 1989 and $830.313 million for 1990).

B = Billion.

Appendix Table 9-5. The Contemporary Canadian Minimum Tax (MT).

Tax Year	Tax Returns Affected	Percent Affected	MT Tax Generated	Source: *RC TS*
Canadian Minimum Tax (MT) Began with the 1986 Tax Year:				
1986	22,080	00.13%	$79.468M	1988, 107
1987	39,550	00.23%	$147.946M	1989, 107
1988	19,340	00.11%	$93.801M	1990, 107
1989	19,330	00.11%	$93.153M	1991, 113
1990	11,930	00.06%	$44.262M	1992, 119
1991	14,610	00.08%	$48.240M	1993, 71
1992	17,220	00.09%	$61.602M	1994, 61
Canadian Minimum Tax Measures Not Separately Disclosed in RC TS Publications:				
1993	Not available	Not available	Not available	
1994	Not available	Not available	Not available	
1995	Not available	Not available	Not available	
1996	Not available	Not available	Not available	

M = Million.

Appendix Table 10-1[1]. An International Comparison of Capital Gains, Estate/Gift, and Income Tax Treatment in 100 Countries – as of January 1, 1991.

Country	Maximum Capital Gains Tax Rates	Maximum Estate/Gift Tax Rates	Maximum Income Tax Rates
1. Argentina	n.a.	n.a.	36%
2. Aruba	n.a.	24%	53.04%
3. Australia	47%	n.a.	47%
4. Austria	25%[2]	60%	50%
5. Bahamas	n.a.	n.a.	n.a.
6. Bangladesh	50%[3]	n.a./20%	50%
7. Barbados	n.a.	n.a.	50%
8. Belgium	55%	80%	55%
9. Bermuda	n.a.	5%/n.a.	n.a.
10. Bolivia	n.a.	20%	10%
11. Botswana	40%	40%	40%
12. Brazil	25%	4%[4]	25%
13. Brunei	n.a.	20%	n.a.
14. Cameroon	60%	10%/20%	60%
15. Canada	36.61%[5]	n.a.	48.81%
16. Cayman Islands	n.a.	n.a.	n.a.
17. Chile	50%	25%	50%
18. China	n.a.	n.a.	45%
19. Columbia	30%	30%	30%
20. Congo	50%	35%	50%
21. Costa Rica	n.a.	n.a.	25%
22. Cyprus	20%	45%	40%
23. Czechoslovakia	55%[6]	n.a.	55%
24. Denmark	50%	50%/90%	68%
25. Dominican Republic	n.a.	32%	82%[7]
26. Ecuador	25%	10%	25%
27. Egypt	40%	30%	65%
28. El Salvador	50%	n.a.	50%
29. Fiji	30%[8]	n.a.	42.5%
30. Finland	39%[9]	32%	59%
31. France	17%	60%	56.8%
32. Gabon	20%	35%	60%
33. Germany	53%	70%	53%
34. Gibraltar	n.a.	25%/n.a.	50%
35. Greece	n.a.	66%	50%
36. Guatemala	34%	25%	34%
37. Guernsey	n.a.	n.a.	20%
38. Guinea	35%	50%	35%
39. Honduras	46%	20%	46%

Appendix Table 10-1[1].   Continued.

Country	Maximum Capital Gains Tax Rates	Maximum Estate/Gift Tax Rates	Maximum Income Tax Rates
40. Hong Kong	n.a.	18%/n.a.	15%
41. Hungary	50%	10%	50%
42. Iceland	39.79%	45%/39.79%	39.79%
43. India	25%[10]	n.a./40%	50%
44. Indonesia	35%	n.a.	35%
45. Iran	75%	75%	75%
46. Ireland	50%	55%	53%
47. Isle of Man	n.a.	n.a.	20%
48. Italy	50%	60%	50%
49. Ivory Coast	60%	45%	60%
50. Jamaica	n.a.	15%/7.5%	33.33%
51. Japan	50%	70%	50%
52. Jersey	n.a.	0.5%	20%
53. Jordan	n.a.	n.a.	45%
54. Kenya	n.a.	n.a.	45%
55. Korea	75%	55%/60%	50%
56. Lesotho	n.a.	n.a.	53%
57. Luxemburg	50%	15%/14.4%	50%
58. Malaysia	20%	10%	35%
59. Mauritius	36.3%	n.a./45%	35%
60. Mexico	35%	n.a.	35%
61. Monaco	n.a.	16%	35%
62. Morocco	15%	4%/15%	52%
63. Namibia	n.a.	n.a.	42%
64. Netherlands	60%	68%	60%
65. Netherlands Antilles	n.a.	24%	57.2%
66. New Zealand	n.a.	40%/25%	33%
67. Niger	n.a.	57%	60%
68. Nigeria	20%	60%	55%
69. Norway	40%	30%	50%
70. Oman	n.a.	n.a.	7.5%
71. Pakistan	45%[11]	n.a.	45%[11]
72. Panama	56%	n.a.	56%
73. Papua New Guinea	n.a.	n.a.	45%
74. Paraguay	30%	36%	30%
75. Peru	n.a.	n.a.[a]	37%
76. Philippines	20%	60%/40%	35%
77. Poland	50%	40%	20%
78. Portugal	20%	50%	40%
79. Puerto Rico	20%	50%	36%
80. Qatar	n.a.	n.a.	n.a.

Appendix Table 10-1[1].   Continued.

Country	Maximum Capital Gains Tax Rates	Maximum Estate/Gift Tax Rates	Maximum Income Tax Rates
81. Saudi Arabia	30%	n.a.	30%
82. Senegal	48%	50%	48%
83. Singapore	n.a.	10%	33%
84. South Africa	n.a.	15%	44%
85. Spain	56%	34%	56%
86. Sri Lanka	40%	n.a.[12]	40%[12]
87. Suriname	50%	15%/n.a.	50%
88. Swaziland	n.a.	n.a.	40%
89. Sweden	30%	65%	50%
90. Switzerland	35%[13]	n.a.	35%[13]
91. Taiwan	40%	60%	40%
92. Thailand	55%	n.a.	55%
93. Trinidad and Tobago	35%	n.a.	35%
94. Turkey	50%	44%	50%
95. United Kingdom	40%	40%	40%
96. United States	28%	55%	31%
97. Uruguay	n.a.	n.a.	30%[14]
98. USSR	n.a.	n.a.	60%
99. Venezuela	45%	55%	45%
100. Zimbabwe	30%	20%/n.a.	60%
Number/Percent without any Capital Gains tax	35		
Number/Percent without any Estate tax		36	
Number/Percent without any Gift tax		38	
Number/Percent with a New Worth tax		9	
Number/Percent without any Income tax			5

[1] As adapted from *Ernst & Young* (1991).

[2] Capital gains are taxed at a maximum of 50% (e.g. 50% exclusion) of the maximum ordinary income tax rate of 50% (e.g. 50% *multiplied by* 50% *equals* 25%). Generally, capital gains are applied to businesses only, and capital gains from privately held assets are not taxed if held for more than one year.

[3] An additional 15% surcharge applies to tax payable, except for nonresident dividend income.

[4] Applied as a transfer tax of 4% on real estate only.

[5] Capital gains are taxed at a maximum of 75% (e.g. 25% exclusion) of the maximum ordinary income tax rate of 48.81% (in the Ontario province). This calculation varies for each of the 10 Canadian provinces.

[6] Capital gains from personal property is tax-exempt if held for more than 2 years.

[7] Includes a flat rate component of 2% to 12%.

Appendix Table 10-1[18].   Continued.

[8] Applies only to gains from undeveloped land.

[9] Capital gain from the taxpayer's personal residence is tax-exempt if held for more than 2 years.

[10] A temporary surcharge of 8% applies above certain levels.

[11] Excludes a 10% surcharge if total income exceeds certain levels. A *Zakat*, a 2.5% tax, applies to Muslim citizens.

[12] A surcharge of 12% is imposed on both income and wealth.

[13] Includes significant municipal taxes to a combined maximum of 35%.

[14] Applied only to agricultural income.

Appendix Table 10-2. Frequency of U.S. Individual Federal Income Tax (FIT) Returns Reporting Net Capital Gain (CG) or Capital Loss (CL) Arising from Sales of Capital Assets.

| Tax Year | Percent with CG or CL | Percent with | | CG/L Ratio | Reference: *IRS SOI* |
		CG	CL		

1922   Return frequency not reported, but short-term (S/T) gains of $742.1 million; long-term (L/T) gains of $249.2 million; and Total gains of $991.4 million v. wages/salaries of $13.7 billion. L/T gains heavily concentrated in high-income ($25 thousand and higher income classes) taxpayer groups and represented 1.00% of "total income".

1922, 9–10

1923   Return frequency not reported, but S/T gains of $866.8 million; L/T gains of $305.4 million; and Total gains of $1.17 billion v. wages/salaries of $14.2 billion. L/T gains heavily concentrated in high-income ($30 thousand and higher income classes) taxpayer groups and represented 1.04% of "total income".

1923, 8–9

1924   Return frequency not reported, but S/T gains of $1.1 billion; L/T gains of $389.1 million; and Total gains of $1.5 billion v. wages/salaries of $13.6 billion. L/T gains heavily concentrated in high-income ($25 thousand and higher income classes) taxpayer groups and represented 1.32% of "total income".

1924, 9

1925   Return frequency not reported, but S/T gains of $2.0 billion; L/T gains of $0.9 billion; and Total gains of $2.9 billion v. wages/salaries of $9.7 billion. L/T gains heavily concentrated in high-income ($30 thousand and higher income classes) taxpayer groups and represented 3.72% of "total income".

1925, 5–7 & 27

1926   Return frequency not reported, but S/T gains of $1.5 billion; L/T gains of $0.9 billion; and Total gains of $2.4 billion v. wages/salaries of $10 billion. L/T gains heavily concentrated in high-income taxpayer groups and represented 3.59% of "total income".

1926, 7

1927   Return frequency reported only for returns in the $5,000 and greater income class. S/T gains of $1.8 billion; L/T gains of $1.1 billion; and Total gains of $2.9 billion v. wages/salaries of $10.2 billion. L/T gains heavily concentrated in high-income taxpayer groups and represented 4.13% of "total income".

| 2.21 CG | 2.21 | Unk | Unk | Unk | For "net income" ≥ $5,000 1927, 4, 7–8 & 10 |

1928   S/T gains of $2.9 billion; L/T gains of $1.9 billion; and Total gains of $4.8 billion v. wages/salaries of $10.9 billion. L/T gains heavily concentrated in high-income taxpayer groups and represented 6.49% of "total income".

1928, 4, 8–9 and 11)

Appendix Table 10-2. Continued.

Tax Year	Percent with CG or CL	Percent with CG	Percent with CL	CG/L Ratio	Reference: *IRS SOI*
	0.68 CG	0.68	Unk	Unk	For all taxpayers 1928, 11
	2.74 CG	2.74	Unk	Unk	For "net income" ≥ $5,000 1928, 4 & 11

For taxpayers with "net income" ≥ $5,000:

Tax Year	Percent with CG or CL	Percent with CG	Percent with CL	CG/L Ratio	Reference: *IRS SOI*
1929	2.59 CG/L	2.29	0.30	7.63	1929, 5 & 12
1930	1.63 CG/L	1.09	0.53	2.06	1930, 5 & 14
1931	1.44 CG/L	0.49	0.95	0.52	1931, 5 & 14
1932	6.06 CG/L	0.57	5.49	0.10	1932, 6 & 14
1933	5.83 CG/L	1.08	4.75	0.23	1933, 6 & 14
1934	33.55 CG/L	14.60	18.95	0.77	1934, 10–11
1935	37.31 CG/L	23.52	13.79	1.71	1935, 10–11
1936	39.25 CG/L	28.92	10.33	2.80	1936, 11–12

For all taxpayers:

Tax Year	Percent with CG or CL	Percent with CG	Percent with CL	CG/L Ratio	Reference: *IRS SOI*
1937	8.83 CG/L	4.51	4.32	1.04	1937, 13–14
1938	3.70 CG/L	1.97	1.73	1.14	1938, 39–40
1939	3.43 CG/L	2.00	1.43	1.40	1939, 26–29
1940	2.11 CG/L	1.29	0.82	1.57	1940, 19–20
1941	1.38 CG/L	0.76	0.62	1.23	1941, 22.24
1942	2.23 CG/L	0.93	1.30	0.72	1942, 32–33
1943	2.45 CG/L	1.54	0.90	1.66	1943, 197 & 199
1944	3.13 CG/L	2.21	0.92	2.41	1944, 159 & 165
1945	4.13 CG/L	3.35	0.78	4.27	1945, 163 & 165
1946	5.20 CG/L	4.25	0.95	4.47	1946, 165 & 167
1947	4.50 CG/L	3.39	1.11	3.06	1947, 169 & 171
1948	4.38 CG/L	1.13	3.25	0.35	1948, 159 & 161
1949	6.90 CG/L	2.78	4.12	0.67	1949, 155 & 157
1950	8.41 CG/L	3.57	4.83	0.74	1950, 113
1951	4.90 CG/L	3.85	1.05	3.67	1951, 29–30
1952	4.78 CG/L	3.60	1.18	3.05	1952, 22
1953	4.80 CG/L	3.44	1.36	2.53	1953, 27
1954	5.42 CG/L	4.25	1.17	3.63	1954, 37
1955	6.10 CG/L	4.98	1.12	4.45	1955, 22
1956	6.64 CG/L	5.32	1.32	4.03	1956, 24
1957	6.64 CG/L	4.91	1.74	2.82	1957, 24
1958	7.43 CG/L	5.87	1.56	3.76	1958, 31
1959	8.14 CG/L	6.65	1.49	4.46	1959, 28
1960	8.19 CG/L	6.29	1.89	3.33	1960, 37

Appendix Table 10-2.   Continued.

Tax Year	Percent with CG or CL	Percent with		CG/L Ratio	Reference: *IRS SOI*
		CG	CL		
1961	9.42 CG/L	7.64	1.78	4.29	1961, 38
1962	9.44 CG/L	6.89	2.55	2.70	1962, 38
1963	10.23 CG/L	7.74	2.49	3.11	1963, 30
1964	10.44 CG/L	8.14	2.30	3.54	1964, 13
1965	10.84 CG/L	8.77	2.07	4.24	1965, 10
1966	10.81 CG/L	8.56	2.26	3.79	1966, 9
1967	11.72 CG/L	9.71	2.01	4.83	1967, 11
1968	12.14 CG/L	10.29	1.85	5.56	1968, 10
1969	12.03 CG/L	9.20	2.84	3.24	1969, 13
1970	10.72 CG/L	7.39	3.33	2.22	1970, 15
1971	9.45 CG/L	6.57	2.87	2.29	1971, 18
1972	11.43 CG/L	8.73	2.70	3.23	1972, 16
1973	10.83 CG/L	7.98	2.86	2.79	1973, 15
1974	9.58 CG/L	6.37	3.21	1.98	1974, 25
1975	9.21 CG/L	6.16	3.06	2.01	1975, 14
1976	10.06 CG/L	7.18	2.88	2.49	1976, 18
1977	8.41 CG/L	6.10	2.31	2.64	1977, 21
1978	9.70 CG/L	7.36	2.35	3.13	1978, 20
1979	9.32 CG/L	7.14	2.19	3.26	1979, 18
1980	9.51 CG/L	7.43	2.08	3.57	1980, 45
1981	9.94 CG/L	7.36	2.58	2.85	1981, 42
1982	10.11 CG/L	7.47	2.64	2.83	1982, 48
1983	9.91 CG/L	7.60	2.31	3.29	1983, 16
1984	10.99 CG/L	8.04	2.95	2.73	1984, 21
1985	10.94 CG/L	8.32	2.62	3.18	1985, 23
1986	12.54 CG/L	10.28	2.25	4.57	1986, 27
1987	14.44 CG/L	10.93	3.51	3.11	1987, 29
1988	11.41 CG/L	7.19	4.22	1.70	1988, 27
1989	11.46 CG/L	7.59	3.87	1.96	1989, 27
1990	10.66 CG/L	6.20	4.46	1.39	1990, 25
1991	11.05 CG/L	7.03	4.02	4.02	1991, 29
1992	11.75 CG/L	7.85	3.90	2.01	1992, 38
1993	12.62 CG/L	8.95	3.67	2.44	1993, 38
1994	14.81 CG/L	7.93	4.84	1.64	1994, 39
1995	15.28 CG/L	8.59	4.34	1.98	1995, 36
1996	16.64 CG/L	9.98	3.84	2.60	1996, 36

Note: Double-digit CPI for 1974 and 1979–1981; stock market "crash" in October, 1987. The capital gains (CGs) reported in this table do not include those CG distributions not reported on the contemporary Form 1040, Schedule D (e.g. CG dividends).

Appendix Table 10-3. Frequency of Canadian Individual Federal Income Tax (FIT) Returns Reporting Capital Gain (CG) or Capital Loss (CL) from ALL SOURCES.

(1972–1996)

Tax Year	Percent with CG or CL	Percent with CG	Percent with CL	CG/L Ratio	Source: *RC TS*
Prior to January 1, 1972, Canada did not Tax Capital Gains:					
1972	2.95	2.22	0.72	3.07	1974, 17 & 163
1973	3.80	2.75	1.05	2.63	1975, 17 & 163
1974	3.37	2.06	1.32	1.56	1976, 17 & 163
1975	2.83	1.68	1.15	1.46	1977, 17 & 163
1976	3.23	2.21	1.03	2.15	1978, 17 & 163
1977	1.54	1.19	0.35	3.45	1979, 17 & 163
1978	4.93	4.17	0.76	5.48	1980, 17 & 157
1979	5.12	4.54	0.58	7.83	1981, 35 & 191
1980	5.33	4.63	0.70	6.59	1982, 35 & 191
1981	5.38	3.93	1.45	2.71	1983, 37 & 213
1982	4.13	2.58	1.55	1.67	1984, 37 & 225
1983	5.81	4.71	1.10	4.28	1985, 37 & 227
1984	6.89	5.57	1.32	4.23	1986, 73 & 268
Capital Gains Inclusion Rate at ONE-HALF or 50% (1985–1987):					
1985	4.31	3.49	0.82	4.24	1987, 74 & 272
Repeal of Deductibility of ≤ $2,000 of Net Capital Losses (1986–):					
1986	5.41	4.74	0.67	7.09	1988, 107 & 278
1987	5.68	4.95	0.73	6.73	1989, 107 & 278
Capital Gains Inclusion Rate Increased to TWO-THIRDS or 67% (1988–1989):					
1988	4.17	3.08	1.08	2.85	1990, 107 & 274
1989	5.42	4.54	0.89	5.12	1991, 113 & 280
Capital Gains Inclusion Rate Increased to THREE-QUARTERS or 75% (1990–):					
1990	4.50	3.37	1.13	2.98	1992, 69 & 286
1991	4.60	3.53	1.07	3.30	1993, 71 & 238
1992	4.98	3.99	0.99	4.01	1994, 61 & 214
1993	6.39	5.61	0.78	7.20	1995, 74 & 141
1994	3.25	2.81	0.44	6.36	1996, 54 & 121
1995	6.45	5.12	1.33	3.85	1997, 72 & 140
1996	8.22	7.27	0.95	7.69	1998, 72 & 139

Appendix Table 10-4.   Frequency of Canadian Individual Federal Income Tax (FIT) Returns Reporting Capital Gain (CG) or Capital Loss (CL) from STOCK SHARES.

(1972–1995)

Tax Year	Percent with CG or CL	Percent with		CG/CL Ratio	Source: Canadian *Taxation Statistics (TS)*
		CG	CL		
Prior to January 1, 1972, Canada did not Tax Capital Gains –					
Reported as Percent with __NET__ CG or CL (TS):					
1972	1.93	Unk	Unk	Unk	1974, 17 & 162
1973	1.89	Unk	Unk	Unk	1975, 17 & 162
1974	1.61	Unk	Unk	Unk	1976, 17 & 162
1975	1.43	Unk	Unk	Unk	1977, 17 & 162
1976	1.53	Unk	Unk	Unk	1978, 17 & 163
1977	0.37	Unk	Unk	Unk	1979, 17 & 162
1978	1.63	Unk	Unk	Unk	1980, 17 & 156
1979	1.87	Unk	Unk	Unk	1981, 35 & 190
1980	2.26	Unk	Unk	Unk	1982, 35 & 190
1981	2.27	Unk	Unk	Unk	1983, 37 & 212
1982	1.94	Unk	Unk	Unk	1984, 37 & 224
1983	2.32	Unk	Unk	Unk	1985, 37 & 226
Reported as Percent with __GROSS__ CG or CL (TS):					
1984	2.13	1.08	1.05	1.02	1986, 73 & 266
Capital Gains Inclusion Rate at __ONE-HALF__ or 50% (1985–1987):					
1985	2.18	1.57	0.61	2.58	1987, 74 & 270
Repeal of Deductibility of ≤ $2,000 of Net Capital Losses (1986–):					
1986	2.43	1.90	0.53	3.58	1988, 107 & 276
1987	2.74	2.14	0.59	3.61	1989, 107 & 276
Capital Gains Inclusion Rate Increased to __TWO-THIRDS__ or 67% (1988–1989):					
1988	2.23	1.29	0.95	1.36	1990, 107 & 272
1989	2.27	1.53	0.75	2.04	1991, 113 & 278
Capital Gains Inclusion Rate Increased to __THREE-QUARTERS__ or 75% (1990–):					
1990	1.66	0.93	0.73	1.27	1992, 69 & 284
1991	1.95	1.20	0.75	1.61	1993, 71 & 236
1992	2.07	1.34	0.73	1.83	1994, 61 & 212
1993	2.75	2.12	0.63	3.36	1995, 74 & 140
1994	1.52	1.27	0.25	5.18	1996, 54 & 120
1995	2.53	1.45	1.08	1.34	1997, 72 & 138
1996	2.91	2.21	0.70	3.16	1998, 72 & 138

Appendix Table 10-5. Long-Term Capital Gain (LTCG) Taxation[1] (1922–) by Revenue Act.

Federal Tax Law	Tax Year(s) for Application	ST & LTCG Holding Period	ST & LTCG Taxation	Source: *IRS SOI*
Rev Acts of 1921, 1924, 1926, 1928, & 1932	1922–33	≤2 yrs (ST) >2 yrs	100 100	1934, 119
Rev Acts of 1934 & 1936	1934–37	≤1 yr (ST) >1 yr to ≤2 yrs >2 yrs to ≤5 yrs >5 yrs to ≤10 yrs >10 yrs	100 80 60 40 30	1942, 324–5
Rev Act of 1938, IRC of 1939 & Rev Act of 1941	1938–41	≤1 yr (ST) >18 mos & ≤2 yrs >2 yrs	100 $66\frac{2}{3}$ 50	1942, 324–5
Rev Act of 1942, Inc Tax Act of 1944 & Rev Act of 1950 & Rev Act of 1951	1942–76	≤6 mo (ST) >6 mo	100 50[2]/100	1942, 324–5; 1950, 330–1; 1951, 157; & 1953, 101
TRA76	1977	≤9 mo (ST) >9 mo	100 100[3,4]	1977, 205
TRA78	1978–6/21/84	≤1 yr (ST) >1 yr	100 40[5]	1978, 3
TRA84	6/22/84–1987	≤6 mo (ST) >6 mo	100 40[5]	1984, 3
TRA86, RRA90, & TRA97	1988-	≤1 yr (ST) >1 yr	100 100[6]	1988, 4 & 1991, 4

[1] Portions of this table were adapted and extended from Cataldo and Savage (2000, 57–58).

[2] Capital gains rates were low and stable; taxpayers were permitted to deduct 50% of LTCG with an alternate tax rate of 25% (1954–1969). Beginning with TRA69 and culminating with TRA76, the alternative tax was restricted and the benefits of the 50% deductions for LTCG were reduced.

[3] Alternative tax rules may have applied (see Chapter 8).

Appendix Table 10-5.   Continued.

[4] The amount of net capital loss available to offset ordinary income in any one year was increased from $1,000 to $2,000 (IRS SOI 1977, vi). TRA76 increased the amount of net capital loss available to offset ordinary income in any one year from $2,000 to $3,000 (IRS SOI 1978, 1), where it remains under current tax law.

[5] TRA78 increased the capital gains deduction to 60% and removed limits on the use of the deduction, lowering the maximum tax rate on LTCG from 49% to 28%. ERTA81 (1981–) retained this 40% feature, further (effectively) lowering the top rate on capital gains to 20% (50% maximum ordinary tax rate *multiplied by* 40%).

[6] LTCG treatment was provided for assets held for more than 6 months *if acquired* before January 1, 1988. TRA86 eliminated the 60% deduction for LTCG, to finance the reduction in ordinary income tax rates, allow the top tax rate to be reduced without providing disproportionate relief to the highest-income group, and to simplify the tax system. The effective maximum tax rate on LTCG was increased from 20% to a flat tax rate of 28% for high-income taxpayers and 33% for taxpayers just below the high-income group. Beginning with 1991, a maximum tax rate of 28% applied to LTCG, compared to a maximum tax rate of 31% on other types of income (IRS SOI 1991, 4). Beginning with the 1997 tax year, adjusted net capital gains were taxed at 20% (10% if the taxpayer is in a 15% bracket). A transitional rule (for realizations occurring between May 6 and July 28, 1997) for mid-term gains provided for a separate tax rate for LTCGs held more than 12 but not more than 18 months.

Appendix Table 11-1. Before-Tax and After-Tax FICA/SECA Rates (1937–2000).

Tax Year(s)	Pre-Tax FICA/ SECA Rates		SECA-Net of Applicable Marginal Tax Rate					
	FICA	SECA	15.0%	28.0%	31.0%	33.0%	36.0%	39.6%
FICA Taxes Imposed (1937–):								
1937–49	1.0%	N/A			N/A			
1950	1.5%	N/A			N/A			
SECA Taxes Imposed (1951–):								
1951–53	1.5%	2.25%			N/A			
1954–56	2.0%	3.0%			N/A			
1957–58	2.25%	3.375%			N/A			
1959	2.5%	3.75%			N/A			
1960–61	3.0%	4.5%			N/A			
1962	3.125%	4.7%			N/A			
1963–65	3.625%	5.4%			N/A			
1966	4.2%	6.15%			N/A			
1967–68	4.4%	6.4%			N/A			
1969–70	4.8%	6.9%			N/A			
1971–72	5.2%	7.5%			N/A			
1973	5.85%	8.0%			N/A			
1974–77	5.85%	7.9%			N/A			
1978	6.05%	8.1%			N/A			
1979–80	6.13%	8.1%			N/A			
1981	6.65%	9.3%			N/A			
1982–83	6.7%	9.35%			N/A			
1984[1]	6.7%[2]	11.3%			N/A			
1985[1]	7.05%	11.8%			N/A			
1986–87[1]	7.15%	12.3%			N/A			
1988–89[1]	7.51%	13.02%			N/A			
SECA Subsidy Phase-Out Completed & SECA Taxes Adjusted/Part Deductible (1990–):								
1990	7.65%	15.3%	13.1%	12.2%	N/A	11.8%	N/A	N/A
1991–92	7.65%	15.3%	13.1%	12.2%	11.9%	N/A	N/A	N/A
1993–2000	7.65%	15.3%	13.1%	12.2%	11.9%	N/A	11.6%	11.3%

[1] Net of credits of 2.7, 2.3, and 2% for 1984, 1985, and the 1986–1989 tax years, respectively.
[2] The employee (employer) FICA rate was at 6.7 (7.0)% for the 1984 tax year. Employer and employee contributions were equivalent for all other tax years.

Appendix Table 11–2. Maximum Employer/Employee & Self-Employed Contributions to FICA/SECA & Hospitalization Insurance/Medicare (1937–2000).

Tax Year(s)	Combined Employer/Employee			Self-Employed		
	FICA	Medicare	Total	SECA	Medicare	Total
1937–1949	$60	$–0–	$60	$–0–	$–0–	$–0–
1950	$90	$–0–	$90	$–0–	$–0–	$–0–
SECA Taxes Imposed (1951-):						
1951–1953	$108	$–0–	$108	$81.00	$–0–	$81.00
1954	$144	$–0–	$144	$108.00	$–0–	$108.00
1955–1956	$168	$–0–	$168	$126.00	$–0–	$126.00
1957–1958	$189	$–0–	$189	$141.75	$–0–	$141.75
1959	$240	$–0–	$240	$180.00	$–0–	$180.00
1960–1961	$288	$–0–	$288	$216.00	$–0–	$216.00
1962	$300	$–0–	$300	$225.60	$–0–	$225.60
1963–1965	$348	$–0–	$348	$259.20	$–0–	$259.20
Hospitalization Insurance/Medicare Component Added to FICA/SECA (1966-):						
Combined Employer/Employee & Self-Employed Contributions to Medicare Become Equivalent (1990-):						
1984	$4,195.80	$982.80	$5,178.60	$3,288.60	SAME	$4,271.40
1985	$4,514.40	$1,069.20	$5,583.60	$3,603.60	SAME	$4,672.80
1986	$4,788.00	$1,218.00	$6,006.00	$3,948.00	SAME	$5,166.00
1987	$4,993.20	$1,270.20	$6,263.40	$4,117.20	SAME	$5,387.40
1988	$5,454.00	$1,305.00	$6,759.00	$4,554.00	SAME	$5,859.00
1989	$5,817.60	$1,392.00	$7,209.60	$4,857.60	SAME	$6,249.60
Combined Employer/Employee & Self-Employed Contributions to FICA & SECA Become Equivalent (1990-):						
1990	$6,361.20	$1,487.70	$7,848.90	SAME	SAME	SAME
1991	$6,621.60	$3,625.00	$10,246.60	SAME	SAME	SAME
1992	$6,882.00	$3,775.80	$10,657.80	SAME	SAME	SAME
1993	$7,142.40	$3,915.00	$11,057.40	SAME	SAME	SAME
Ceiling for Hospitalization Insurance/Medicare Component Removed (1994-):						
1994	$7,514.40	NO CEILING		SAME	SAME	SAME
1995	$7,588.80	NO CEILING		SAME	SAME	SAME
1996	$7,774.80	NO CEILING		SAME	SAME	SAME
1997	$8,109.60	NO CEILING		SAME	SAME	SAME
1998	$8,481.60	NO CEILING		SAME	SAME	SAME
1999	$9,002.40	NO CEILING		SAME	SAME	SAME
2000	$9,448.80	NO CEILING		SAME	SAME	SAME

Appendix Table 11-3. Additional Descriptive Measures and Variables Related to the SECA Tax Subsidy (1951–1989).

Tax Year(s)	Weighted Average AGI		SE Profit Percentage	Percent w/SchC&F	EIC	IRC 179
	All Returns	SE Only				
1951	$3,676	$4,209	85.398%	12.941%	–0–	$–0–
1952	$3,867	$4,413	84.273%	12.158%	–0–	$–0–
1953	$4,022	$4,291	82.691%	12.799%	–0–	$–0–
1954	$4,107	$4,441	81.187%	13.72%	–0–	$–0–
1955	$4,334	$4,580	81.702%	14.155%	–0–	$–0–
1956	$4,612	$4,936	82.264%	15.157%	–0–	$–0–
1957	$4,786	$5,106	82.122%	13.791%	–0–	$–0–
1958	$4,865	$5,173	82.103%	14.184%	–0–	$3,245
1959	$5,189	$5,582	80.080%	14.285%	–0–	$3,236
1960	$5,293	$5,628	79.445%	14.09%	–0–	$3,233
1961	$5,451	$5,992	80.153%	14.16%	–0–	$3,192
1962	$5,653	$6,335	80.180%	13.757%	–0–	$3,183
1963	$5,870	$6,655	77.962%	13.928%	–0–	$3,181
1964	$6,385	$7,563	77.386%	14.048%	–0–	$3,185
1965	$6,699	$8,338	78.080%	13.048%	–0–	$3,163
1966	$6,995	$8,909	78.135%	12.709%	–0–	$3,138
1967	$7,405	$9,615	76.457%	12.56%	–0–	$3,153
1968	$7,729	$10,193	75.078%	12.167%	–0–	$3,122
1969	$8,176	$11,084	76.630%	11.722%	–0–	$3,119
1970	$8,685	$11,090	72.057%	12.367%	–0–	$3,141
1971	$9,234	$12,865	70.168%	12.292%	–0–	$3,146
1972	$9,854	$13,048	71.455%	12.219%	–0–	$3,111
1973	$10,528	$14,622	71.402%	12.123%	–0–	$3,082
1974	$11,192	$15,323	67.934%	11.98%	–0–	$3,061
1975	$11,912	$15,926	67.770%	12.158%	–0–	$3,074
1976	$12,893	$17,239	68.353%	12.244%	10%	$3,050
1977	$13,892	$18,983	68.516%	12.064%	10%	$3,016
1978	$15,127	$21,007	68.621%	12.141%	10%	$2,991
1979	$16,187	$22,307	67.577%	12.049%	10%	$2,968
1980	$17,628	$23,801	64.658%	12.236%	10%	$2,964
1981	$19,145	$25,216	61.561%	12.802%	10%	$–0–
1982	$20,036	$25,236	60.199%	13.407%	10%	$5K
1983	$20,806	$26,985	62.414%	13.904%	10%	$5K
1984	$22,210	$28,950	63.663%	14.01%	10%	$5K
1985	$23,481	$30,858	65.644%	14.284%	11%	$5K
1986	$24,838	$33,899	67.637%	14.445%	11%	$5K
1987	$26,703	$36,739	70.924%	14.414%	14%	$10K
1988	$28,466	$40,121	72.052%	14.528%	14%	$10K
1989	$29,656	$41,712	72.921%	14.733%	14%	$10K

K = 000 omitted.

Appendix Table 11-4. Descriptive Measures and Variables Related to the SECA Tax Subsidy (1951–1989).

Tax Year(s)	FICA/SECA Wage Base	FICA *less* SECA Rate	SECA Tax Subsidy Nominal	SECA Tax Subsidy Adjusted	CPI
1951	$3,600	0.75%	$27	$159	7.93%
1952	$3,600	0.75%	$27	$156	2.151%
1953	$3,600	0.75%	$27	$154	0.804%
1954	$3,600	1.0%	$36	$205	0.484%
1955	$4,200	1.0%	$42	$240	−0.402%
1956	$4,200	1.0%	$42	$236	1.468%
1957	$4,200	1.125%	$47.25	$257	3.635%
1958	$4,200	1.125%	$47.25	$250	2.661%
1959	$4,800	1.25%	$60.00	$315	0.876%
1960	$4,800	1.5%	$72.00	$372	1.601%
1961	$4,800	1.5%	$72.00	$368	0.988%
1962	$4,800	1.55%	$74.40	$376	1.09%
1963	$4,800	1.85%	$88.80	$444	1.195%
1964	$4,800	1.85%	$88.80	$438	1.398%
1965	$4,800	1.85%	$88.80	$430	1.705%
1966	$6,600	2.25%	$148.50	$700	2.792%
1967	$6,600	2.4%	$158.40	$725	2.907%
1968	$7,800	2.4%	$187.20	$823	4.177%
1969	$7,800	2.7%	$210.60	$878	5.392%
1970	$7,800	2.7%	$210.60	$829	5.905%
1971	$7,800	2.9%	$226.20	$854	4.379%
1972	$9,000	2.9%	$261.00	$955	3.137%
1973	$10,800	3.7%	$399.60	$1,376	6.22%
1974	$13,200	3.8%	$501.60	$1,557	10.942%
1975	$14,100	3.8%	$535.80	$1,524	9.145%
1976	$15,300	3.8%	$581.40	$1,563	5.805%
1977	$16,500	3.8%	$627.00	$1,582	6.547%
1978	$17,700	4.0%	$708.00	$1,660	7.64%
1979	$22,900	4.16%	$952.64	$2,012	11.014%
1980	$25,900	4.16%	$1,077.44	$2,003	13.58%
1981	$29,700	4.0%	$1,188.00	$1,996	10.656%
1982	$32,400	4.05%	$1,312.20	$2,078	6.087%
1983	$35,700	4.05%	$1,445.85	$2,219	3.19%
1984	$37,800	2.1%	$793.80	$1,334	4.37%
1985	$39,600	2.3%	$910.80	$1,293	3.556%
1986	$42,000	2.0%	$840.00	$1,173	1.643%
1987	$43,800	2.0%	$876.00	$1,180	3.75%
1988	$45,000	2.0%	$900.00	$1,165	4.019%
1989	$48,000	2.0%	$960.00	$1,185	4.833%

[1] The FICA wage base was stable at $3,000 for the 1937–1950 tax years. For the post–1989 period, it increased to $51,300 (1990), $53,400 (1991), $55,500 (1992), $57,600 (1993), $60,600 (1994), $61,200 (1995), $62,700 (1996), $65,400 (1997), $68,400 (1998), and $72,600 (1999).

Appendix Table 11-5. Hospital Insurance (Medicare) Rates – Included in FICA & SECA (1966–2000).

Tax Year(s)	Employee Medicare	Combined Employer & Employee Medicare	Less: SECA Medicare	Equals: SECA Subsidy	Wage Base
1966	0.35%	0.7%	0.35%	0.35%	$6,600
1967	0.5%	1.0%	0.5%	0.5%	$6,600
1968–1971	0.6%	1.2%	0.6%	0.6%	$7,800
1972	0.6%	1.2%	0.6%	0.6%	$9,000
1973	1.0%	2.0%	1.0%	1.0%	$10,800
1974	0.9%	1.8%	0.9%	0.9%	$13,200
1975	0.9%	1.8%	0.9%	0.9%	$14,100
1976	0.9%	1.8%	0.9%	0.9%	$15,300
1977	0.9%	1.8%	0.9%	0.9%	$16,500
1978	1.0%	2.0%	1.0%	1.0%	$17,700
1979	1.05%	2.1%	1.05%	1.05%	$22,900
1980	1.05%	2.1%	1.05%	1.05%	$25,900
1981	1.3%	2.6%	1.3%	1.3%	$29,700
1982	1.3%	2.6%	1.3%	1.3%	$32,400
1983	1.3%	2.6%	1.3%	1.3%	$35,700

SECA Hospital Insurance (Medicare) Tax Subsidy Eliminated (1984–):

Tax Year(s)	Employee Medicare	Combined Employer & Employee Medicare	Less: SECA Medicare	Equals: SECA Subsidy	Wage Base
1984	1.3%	2.6%	2.6%	–0–%	$37,800
1985	1.35%	2.7%	2.7%	–0–%	$39,600
1986	1.45%	2.9%	2.9%	–0–%	$42,000
1987	1.45%	2.9%	2.9%	–0–%	$43,800
1988	1.45%	2.9%	2.9%	–0–%	$45,000
1989	1.45%	2.9%	2.9%	–0–%	$48,000
1990	1.45%	2.9%	2.9%	–0–%	$51,300
1991	1.45%	2.9%	2.9%	–0–%	$125,000
1992	1.45%	2.9%	2.9%	–0–%	$130,200
1993	1.45%	2.9%	2.9%	–0–%	$135,000
1994-	1.45%	2.9%	2.9%	–0–%	NO CEILING

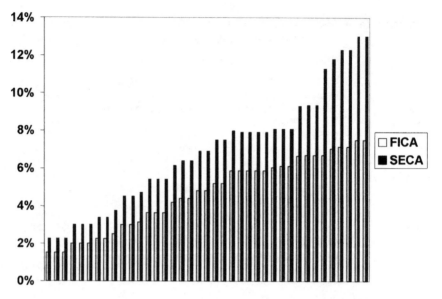

Appendix Fig. 11-1. FICA & SECA Tax Rates (1951–1989).

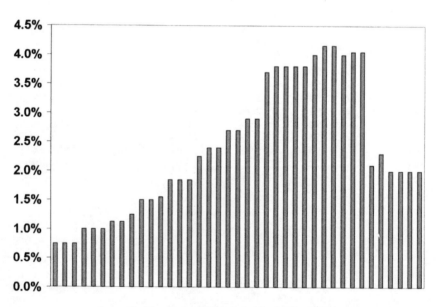

Appendix Fig. 11-2. SECA Subsidy (1951–1989).

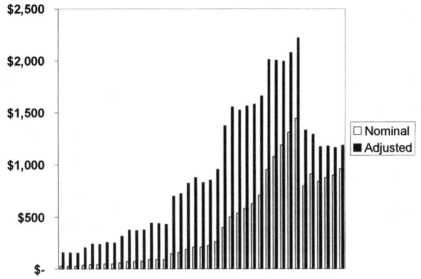

Appendix Fig. 11-3. Nominal & CPI-Adjusted (1995 dollars) SECA Subsidy (1951–1989).

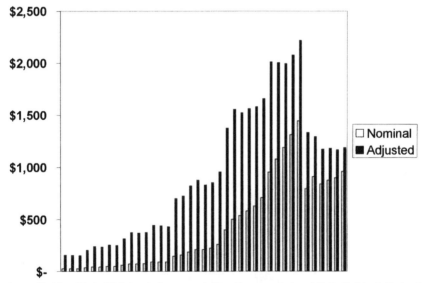

Appendix Fig. 11-4. Weighted Average AGIs Compared for All Individual Federal Income Tax Returns and those Individual Federal Income Tax Returns with SECA Tax (1951–1989).

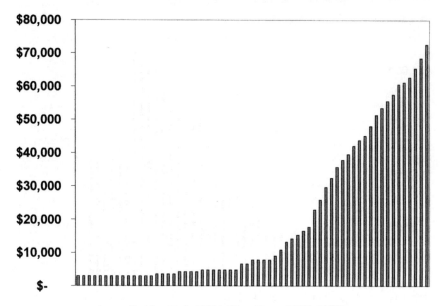

Appendix Fig. 11-5. FICA Wage Base (1937–1999).

Appendix Table 12-1. A Summary of U.S. Individual Retirement Account (IRA) Contributions for Primary (P), Secondary (S), and Combined (C) Taxpayers (1975–1996).

Tax Year	Tax Returns Affected	Percent Affected	Aggregate IRA Contributions	IRA Contributions as a Percent of AGI	Source: *IRS SOI*
1975	1,211,794C	1.47%	$1.436B	0.15%	1989, 2
1976	1,640,975C	1.94%	$1.968B	0.19%	1976, 32
1977	2,002,833C	2.31%	$2.458B	0.21%	1977, 24
1978	2,382,741C	2.65%	$2.970B	0.23%	1978, 22
1979	2,451,955C	2.65%	$3.199B	0.22%	1979, 22
1980	2,564,421C	2.73%	$3.431B	0.21%	1990, 2
1981	3,415,053C	3.58%	$4.750B	0.27%	1981, 45

All Primary (P) Taxpayers, Including Qualified Plan Active Participants, May Contribute ≤ $2,000 (Increased from ≤ $1,500 Before the 1982 Tax Year) to an IRA (1982–):

1982	12,010,038C	12.60%	$28.274B	1.53%	1982, 52
1983	13,613,167C	14.13%	$32.061B	1.65%	1984, 2
1984	15,232,856C	15.32%	$35.374B	1.65%	1985, 2
1985	16,205,846C	16.94%	$38.212B	1.66%	1991, 2
1986	15,535,531C	15.08%	$37.759B	1.52%	1987, 2

IRA Deductions are Curtailed for Taxpayers who are Active Participants in a Qualified Plan or Exceed Specified Filing Status-Based AGI Thresholds (1987–):

1987	**7,318,727C**	6.84%	$14.066B	0.51%	1988, 2
1988	5,825,108P	5.31%	$8.581B	0.28%	1992, 3
	2,615,832S	2.38%	$3.301B	0.11%	1992, 3
	8,440,940C	7.69%	$11.882B	0.39%	1992, 3
1989	5,280,531P	4.71%	$7.807B	0.24%	1993, 3
	2,372,984S	2.12%	$3.022B	0.09%	1993, 3
	7,653,515C	6.83%	$10.829B	0.33%	1993, 3
1990	4,761,622P	4.19%	$7.128B	0.21%	1993, 3
	2,106,852S	1.85%	$2.730B	0.08%	1993, 3
	6,868,474C	6.04%	$9.858B	0.29%	1993, 3
1991	4,211,741P	3.67%	$6.401B	0.18%	1993, 3
	1,912,395S	1.67%	$2.629B	0.08%	1993, 3
	6,124,136C	5.34%	$9.030B	0.26%	1993, 3
1992	4,036,901P	3.55%	$6.192B	0.17%	1993, 3
	1,837,901S	1.62%	$2.504B	0.07%	1993, 3
	5,873,986C	5.17%	$8.696B	0.24%	1993, 3

Appendix Table 12-1. A Summary of U.S. Individual Retirement Account (IRA) Contributions for Primary (P), Secondary (S), and Combined (C) Taxpayers (1975–1996).

Tax Year	Tax Returns Affected	Percent Affected	Aggregate IRA Contributions	IRA Contributions as a Percent of AGI	Source: *IRS SOI*
1993	3,998,976P	3.49%	$6.050B	0.16%	1993, 3
	1,799,919S	1.57%	$2.477B	0.07%	1993, 3
	5,798,895C	5.06%	$8.527B	0.23%	1993, 3
1994	3,902,183P	3.37%	$5.974B	0.15%	1996, 3
	1,707,974S	1.47%	$2.415B	0.06%	1996, 3
	5,610,157C	4.84%	$8.389B	0.21%	1996, 3
1995	3,889,574P	3.29%	$5.945B	0.14%	1996, 3
	1,661,103S	1.41%	$2.393B	0.06%	1996, 3
	5,550,677C	4.70%	$8.338B	0.20%	1996, 3
1996	3,970,788P	3.30%	$6.207B	0.14%	1996, 3
	1,663,101S	1.38%	$2.420B	0.05%	1996, 3
	5,633,889C	4.68%	$8.628B	0.19%	1996, 3

Appendix Table 12-2. A Summary of Canadian Registered Retirement Savings Plan (RRSP) Contributions (1958–1995).

Tax Year	Tax Returns Affected	Percent Affected	Aggregate RRSP Contributions	RRSP Contributions as a Percent of TI	Source: *DNR* or *RC TS*
"Registered Retirement Savings Plan Contributions" First Reported					
Separately from Employer Pension Plan Measures (1958–):					
1958	Unknown	Unknown	$0.019004B	0.23%	1960, 39
1959[1]	Unknown	Unknown	Unknown	Unknown	1961, 26
1960	Unknown	Unknown	$0.027526B	0.28%	1962, 39
1961	Unknown	Unknown	$0.034322B	0.33%	1963, 41
1962	Unknown	Unknown	$0.040057B	0.36%	1964, 31
1963	Unknown	Unknown	$0.045993B	0.38%	1965, 33
1964	Unknown	Unknown	$0.057321B	0.40%	1966, 23
1965	Unknown	Unknown	$0.081403B	0.49%	1967, 25
1966	Unknown	Unknown	$0.100618B	0.52%	1968, 35
1967	Unknown	Unknown	$0.118864B	0.53%	1969, 35
1968	171,894	2.02%	$0.142618B	0.55%	1970, 35 & 145
1969	205,879	2.32%	$0.178600B	0.60%	1971, 37 & 149
1970	248,719	2.71%	$0.225214B	0.67%	1972, 39 & 151
1971	347,674	3.65%	$0.319779B	0.85%	1973, 39 & 151
1972	545,416	6.75%	$0.645123B	1.65%	1974, 39 & 153
1973	757,925	6.89%	$0.922595B	1.95%	1975, 39 & 153
1974	936,385	10.49%	$1.243724B	2.13%	1976, 14 & 153
1975	1,078,152	12.70%	$1.524281B	2.22%	1977, 13 & 153
1976	1,291,349	10.46%	$2.115539B	2.69%	1978, 17 & 153
1977	1,425,239	11.32%	$2.368901B	2.76%	1979, 13 & 153
1978	1,571,174	10.97%	$2.675385B	2.76%	1980, 39 & 147
1979	1,725,959	11.76%	$3.091044B	2.84%	1981, 57
1980	1,916,372	12.98%	$3.676073B	2.90%	1982, 57
1981	1,954,002	12.87%	$3.879173B	2.62%	1983, 59
1982	2,100,333	13.80%	$4.317349B	2.69%	1984, 61
1983	2,329,201	15.22%	$4.997187B	3.07%	1985, 61
1984	2,645,080	17.01%	$5.791294B	3.31%	1986, 97
1985	2,892,892	18.24%	$6.671534B	3.53%	1987, 101
1986	3,216,350	19.45%	$7.919749B	3.92%	1988, 107
1987	3,483,650	20.41%	$9.024445B	4.11%	1989, 107
1988	3,802,260	21.63%	$10.599878B	2.94%	1990, 107
1989	4,161,450	22.95%	$11.937592B	3.02%	1991, 113
1990	4,139,920	22.07%	$10.626239B	2.53%	1992, 119
1991	4,617,640	24.24%	$13.370792B	3.13%	1993, 71
1992	4,836,410	24.88%	$14.784369B	3.42%	1994, 61
1993	5,132,280	25.88%	$17.499790B	4.02%	1995, 74

Appendix Table 12-2.   Continued.

Tax Year	Tax Returns Affected	Percent Affected	Aggregate RRSP Contributions	RRSP Contributions as a Percent of TI	Source: DNR or RC TS
1994	5,367,570	26.63%	$19.284725B	4.34%	1996, 54
1995	5,727,650	27.92%	$21.162754B	4.57%	1997, 72
1996	6,000,000	28.84%	$23.756506B	4.94%	1998, 72

[1] Combined with employer pension fund contributions. Separate ("RRSP" and "RRSP Premium" predecessor) measures are not available.

Appendix Table 12-3.   A Summary of U.S. Contributions to a Keogh or Simplified Employee Pension (SEP) Retirement Plans.

Tax Year	Tax Returns Affected	Percent Affected	Aggregate Keogh & SEP Contributions	Keogh & SEP Contributions as a Percent of AGI	Source: IRS SOI
1975	595,852	0.72%	$1.604B	0.17%	1989, 2
1976	Unknown	Unknown	Unknown	Unknown	
1977	576,982	0.67%	$1.835B	0.16%	1977, 24
1978	627,367	0.70%	$1.994B	0.15%	1978, 22
1979	590,189	0.64%	$2.029B	0.14%	1979, 22
1980	568,936	0.61%	$2.008B	0.12%	1990, 2
1981	557,038	0.58%	$2.012B	0.11%	1981, 46
1982	559,011	0.59%	$2.483B	0.13%	1983, 53
1983	656,038	0.68%	$2.938B	0.15%	1984, 2
1984	648,958	0.65%	$4.072B	0.19%	1985, 2
1985	675,822	0.66%	$5.182B	0.22%	1990, 2
1986	773,926	0.75%	$6.195B	0.25%	1987, 2
1987	675,822	0.66%	$5.182B	0.22%	1991, 2
1988	814,586	0.74%	$6.627B	0.21%	1992, 3
1989	822,353	0.73%	$6.326B	0.19%	1993, 3
1990	824,327	0.72%	$6.778B	0.20%	1993, 3
1991	840,087	0.73%	$6.913B	0.20%	1993, 3
1992	919,187	0.81%	$7.592B	0.21%	1993, 3
1993	947,949	0.83%	$8.160B	0.22%	1993, 3
1994	995,844	0.86%	$8.195B	0.21%	1996, 3
1995	1,032,102	0.87%	$8.734B	0.21%	1996, 3
1996	1,079,413	0.90%	$8.979B	0.20%	1996, 3

Appendix Table 13-1.   Descriptive Statistics of U.S. Estate Tax Returns[1].

Tax Year	Tax Rates[2]		Number of Returns	Unified Credit	Source
	Min	Max			
1916[3]	1.0%	10.0%		N/A	*IRS SOI* 1949, 406 and 456–7
1917[4]	2.0%	25.0%		N/A	*IRS SOI* 1949, 406 and 456–7
1918	1.0%	25.0%	45,126[5]	N/A	*IRS SOI* 1949, 406 and 456–7
1919	1.0%	25.0%		N/A	*IRS SOI* 1949, 406 and 456–7
1920	1.0%	25.0%		N/A	*IRS SOI* 1949, 406 and 456–7
1921	1.0%	25.0%		N/A	*IRS SOI* 1949, 406 and 456–7
1922	1.0%	25.0%	13,876[6]	N/A	*IRS SOI* 1949, 406 and 456–7
1923	1.0%	25.0%	15,119	N/A	*IRS SOI* 1949, 406 and 456–7
1924[7]	1.0%	40.0%	14,513	N/A	*IRS SOI* 1949, 406 and 456–7
1925	1.0%	40.0%	16,019	N/A	*IRS SOI* 1949, 406 and 456–7
1926[8]	1.0%	20.0%	14,567	N/A	*IRS SOI* 1949, 406 and 456–7
1927	1.0%	20.0%	10,700	N/A	*IRS SOI* 1949, 406 and 456–7
1928	1.0%	20.0%	10,236	N/A	*IRS SOI* 1949, 406 and 456–7
1929	1.0%	20.0%	10,343	N/A	*IRS SOI* 1949, 406 and 456–7
1930	1.0%	20.0%	10,382	N/A	*IRS SOI* 1949, 406 and 456–7
1931	1.0%	20.0%	9,889	N/A	*IRS SOI* 1949, 406 and 456–7
1932[9]	2.0%	65.0%	8,507	N/A	*IRS SOI* 1949, 406 and 456–7
1933	2.0%	65.0%	10,275	N/A	*IRS SOI* 1949, 406 and 456–7
1934[10]	2.0%	80.0%	11,853	N/A	*IRS SOI* 1949, 406 and 456–7
1935	3.0%	90.0%	12,724	N/A	*IRS SOI* 1949, 406 and 456–7
1936	3.0%	90.0%	13,321	N/A	*IRS SOI* 1949, 406 and 456–7
1937	3.0%	90.0%	17,032	N/A	*IRS SOI* 1949, 406 and 456–7
1938	3.0%	90.0%	17,642	N/A	*IRS SOI* 1949, 406 and 456–7
1939	3.0%	90.0%	16,926	N/A	*IRS SOI* 1949, 406 and 456–7
1940[11]	3.0%	90.0%	16,876	N/A	*IRS SOI* 1949, 406 and 456–7
1941	4.0%	97.0%	17,122	N/A	*IRS SOI* 1949, 406 and 456–7
1942[12]	4.0%	97.0%	17,396	N/A	*IRS SOI* 1949, 406 and 456–7
1943	4.0%	97.0%	16,033	N/A	*IRS SOI* 1949, 406 and 456–7
1944	4.0%	97.0%	14,857	N/A	*IRS SOI* 1949, 406 and 456–7
1945	4.0%	97.0%	16,550	N/A	*IRS SOI* 1949, 406 and 456–7
1946	4.0%	97.0%	Not available	N/A	Not available
1947	4.0%	97.0%	22,007	N/A	*IRS SOI* 1949, 406 and 456–7
1948	4.0%	97.0%	24,381	N/A	*IRS SOI* 1949, 406 and 456–7
1949	4.0%	97.0%	25,904	N/A	*IRS SOI* 1949, 406 and 456–7
1950	4.0%	97.0%	27,144	N/A	*IRS SOI* 1949, 406 and 456–7

Intermittent reporting for the post–1950 tax years:

Tax Year	Min	Max	Number of Returns	Unified Credit	Source
1951[13]	4.0%	97.0%	29,002	N/A	*Fiduciary/Gift/Estate IRS SOI* 1965, 57
1954[14]	3.0%	77.0%	37,672	N/A	*Estate IRS SOI* 1954, 6
1955	3.0%	77.0%	37,565[15]	N/A	*Estate IRS SOI* 1954, 5–6

Appendix Table 13-1.   Continued.

Tax Year	Tax Rates[2]		Number of Returns	Unified Credit	Source
	Min	Max			
1957	3.0%	77.0%	47,381	N/A	*Estate/Gift IRS SOI* 1956, 3
1959	3.0%	77.0%	56,977	N/A	*Fiduciary/Gift/Estate IRS SOI* 1958, 51
1961	3.0%	77.0%	65,789	N/A	*SOI* 1958, 51, 1962, 51
1963	3.0%	77.0%	79,743	N/A	*SOI* 1958, 51, 1962, 51
1966	3.0%	77.0%	98,905	N/A	*SOI* 1958, 51, 1965, 57
1970	3.0%	77.0%	133,944	N/A	*Estate IRS SOI* 1969, 2
1973	3.0%	77.0%	175,000Est.	N/A	*Estate IRS SOI* 1976, 2
1977	18.0%	70.0%	200,747	N/A[16]	*Estate IRS SOI* 1976, 11
1982	18.0%	65.0%	Not available	$62,800	Not available
1983	18.0%	60.0%	Not available	$79,300	Not available
1984	18.0%	55.0%	Not available	$96,300	Not available
1985	18.0%	55.0%	Not available	$121,800	Not available
1986	18.0%	55.0%	Not available	$155,800	Not available
1987	18.0%	55.0%	Not available	$192,800	Not available
1989	18.0%	55.0%	50,434	$192,800	*SOI Bulletin* 1996–97, 9
1992	18.0%	55.0%	60,082	$192,800	*SOI Bulletin* 1996–97, 9
1993	18.0%	55.0%	60,211	$192,800	*SOI Bulletin* 1995, 101
1994	18.0%	55.0%	68,595	$192,800	*SOI Bulletin* 1996–97, 8
1995	18.0%	55.0%	69,772	$192,800	*SOI Bulletin* 1996–97, 8
1997	18.0%	55.0%	102,000	$192,800	*SOI Bulletin* 1998–99, 184
1998	18.0%	55.0%	114,000Est.	$202,050[17]	*SOI Bulletin* 1998–99, 184
1999	18.0%	55.0%	125,000Proj.	$211,300	*SOI Bulletin* 1998–99, 184
2000	18.0%	55.0%	137,000Proj.	$220,550	*SOI Bulletin* 1998–99, 184
2001	18.0%	55.0%	148,000Proj.	$220,550	*SOI Bulletin* 1998–99, 184
2002	18.0%	55.0%	160,000Proj.	$229,800	*SOI Bulletin* 1998–99, 184
2003	18.0%	55.0%	171,000Proj.	$229,800	*SOI Bulletin* 1998–99, 184
2004	18.0%	55.0%	183,000Proj.	$287,300	*SOI Bulletin* 1998–99, 184
2005	18.0%	55.0%	195,000Proj.	$326,300	*SOI Bulletin* 1998–99, 184
2006–	18.0%	55.0%	Not available	$345,800	

[1] Much of the information contained in this table has been adapted from IRS SOI Bulletin 1996–97 (11).

[2] Includes the impact of both basic and supplemental taxes imposed on lowered exemption amounts for the 1932–1953 tax years.

[3] Effective period of the Revenue Acts of 1916, 1917, 1918 and 1921 was September 9th, 1916–4 p.m., June 2, 1924. Exemptions of $50,000 (1916–1925).

[4] Minimum (maximum) tax rates at 1.5% (15%) from March 3rd–October 3rd.

[5] Returns filed from September 9, 1916–January 15, 1922.

[6] Returns filed from January 16–December 31, 1922. Returns for the calendar year approximated 14,150.

Appendix Table 13-1. Continued.

[7] Effective period of the Revenue Act of 1924 was 4:01 p.m., June 2, 1924–10:24 a.m., February 26, 1926.

[8] Effective period of the Revenue Act of 1926 was 10:25 a.m., February 26, 1926–4:59 p.m., June 6, 1932. Exemption increased from $50,000 to $100,000 (1926–1953).

[9] Effective period of the Revenue Act of 1926 (as amended) and the IRC was 5:00 p.m., June 6, 1932 and thereafter. Effective period for the Revenue Act of 1932 was also 5:00 p.m., June 6, 1932, but through May 10, 1934. For the 1932–1953 tax years a *supplemental tax* was imposed (1932–1934 at 1%; 1935–1940 at 2%; and 1941–1953 at 3%).

[10] Effective period of the Revenue Act of 1932 (as amended by the Revenue Act of 1934) was May 11, 1934–August 30, 1935.

[11] Effective period of the Revenue Act of 1932 (as amended by the Revenue Act of 1935), the IRC, and the 1939, 1940, and 1941 amendments to the IRC was August 31, 1935–October 21, 1942. For the 1940 tax year a 10% was surtax was added.

[12] Effective period of the Revenue Act of 1942 amendments to the IRC was October 21, 1942–October 20, 1951.

[13] Effective period of the Revenue Act of 1951 was October 21, 1951–August 16, 1954.

[14] Effective period of the IRC of 1954 was August 17, 1954 and thereafter.

[15] These returns were classified according to the Revenue Act filed under: 13,635 under the 1954 IRC (date of death on or after August 17, 1954), 22,905 under the 1948 Revenue Act (date of death from January 1, 1948–August 16, 1954), and 55 under 1942 and prior Revenue Acts (date of death before January 1, 1948) (Estate IRS SOI 1954, 9). Later IRS SOI issues may not provide the composition of returns filed.

[16] Beginning with the 1997 tax year, the exemption was replaced by the unified credit (below). However, exemption equivalents increased from $60,000 (1954–1976) to $120,000 (1977), to $134,000 (1978), to $147,000 (1979), to $161,000 (1980), to $175,000 (1981), to $225,000 (1982), to $275,000 (1983), to $325,000 (1984), to $400,000 (1985), to $500,000 (1986), and to $600,000 (1987–1997), when graduated tax rates and the unified credit was phased-out for estates valued in excess of $10 million.

[17] Exemption equivalents increased (or are scheduled to increase) from $600,000 to $625,000 (1998), to $650,000 (1999), to $675,000 (2000), to $700,000 (2002), to $850,000 (2004), $950,000 (2005), and to $1 million (2006–).

Appendix Table 13-2.   Descriptive Statistics from U.S. Gift Tax Returns.

Tax Year[1]	Total Returns	Percent Taxable	Net/Taxable Gifts[2]	Gift Tax[3]	Source: IRS SOI
1932[4]	1,747	14.02%	$0.02	$1.11	1948, 424
1933	3,683	23.84%	$0.10	$8.94	1949, 430
1934	9,270	27.27%	$0.54	$68.38	1949, 430
1935	22,563	38.64%	$1.20	$162.80	1949, 430
1936[5]	13,420	28.09%	$0.13	$15.66	1949, 430
1937	13,695	30.14%	$0.18	$22.76	1949, 430
1938	11,042	31.83%	$0.14	$17.84	1949, 430
1939[6]	12,226	32.14%	$0.13	$18.70	1949, 430
1940	15,623	31.56%	$0.23	$34.45	1949, 430
1941	25,788	34.67%	$0.48	$69.82	1949, 430
1942	16,906	25.91%	$0.12	$24.67	1949, 430
1943[7]	16,987	27.41%	$0.12	$29.64	1949, 430
1944	18,397	27.06%	$0.15	$37.78	1949, 430
1945	20,095	27.57%	$0.17	$36.63	1949, 430
1946	24,826	27.42%	$0.27	$62.34	1949, 430
1947	24,857	27.44%	$0.26	$64.40	1949, 430
1948	26,200	25.03%	$0.21	$45.34	1949, 430
1949	31,547	19.38%	$0.18	$36.09	1949, 430
1950	39,056	21.42%	$0.34	$77.61	1951, 154
1951[8]	41,703	20.05%	$0.30	$67.43	1951, 154

[1] For the post–1996 tax years, estimated and projected gift tax returns are 256,000 (1997), 261,000 (1998), 277,000 (1999), 292,000 (2000), 304,000 (2001), 318,000 (2002), 331,000 (2003), 344,000 (2004), and 358,000 (2005) (IRS SOI 1998–99, 184).
[2] In billions of dollars.
[3] In millions of dollars.
[4] From June 7–December 31, 1932. The annual exclusion was $5,000 and specific lifetime exemptions at $50,000 (1932–1935).
[5] Specific lifetime exemptions decreased from $50,000 to $40,000 (1936–1938).
[6] The annual exclusion decreased from $5,000 to $4,000 (1939–1942).
[7] The annual exclusion decreased from $4,000 to $3,000 and specific lifetime exemptions decrease from $40,000 to $30,000 (1943–1962).
[8] Intermittent reporting for the post–1951 tax years.

Appendix Table 13-3. Descriptive Statistics from Canadian Estate Tax
Returns.

Tax Year	Tax Revenues[1]	Number of Returns[2]	Total Net Assets	Percent Stocks	Source
1941[3]	$ ≈ 0.800M	1,488	$23.965M	Unknown	*TS* 1947, 15; 1948, 161; 1949, 161; 1955, 141; & 1957, 15
1942	$6.957M	7,298	$145.197M	Unknown	*TS* 1947, 15; 1948, 161; 1949, 161; 1955, 141; & 1957, 15
1943	$13.273M	9,348	$225.093M	Unknown	*TS* 1947, 15; 1948, 161; 1949, 161; 1955, 141; & 1957, 15
1944	$15.020M	10,478	$290.653M	Unknown	*TS* 1947, 15; 1948, 161; 1949, 161; 1955, 141; &. 1957, 15
1945	$17.251M	10,162	$304,173M	Unknown	*TS* 1947, 15; 1948, 161; 1949, 161; 1955, 141; & 1957, 15
1946	$21.448M	12,351	$374,812M	Unknown	*TS* 1947, 15; 1948, 161; 1949, 161; 1955, 141; & 1957, 15
1947	$23.576M	12,442	$375,598M	Unknown	*TS* 1947, 15; 1948, 161; 1949, 161; 1955, 141; & 1957, 15
1948[4]	$30.828M	5,159	$289,686M	Unknown	*TS* 1950, 157; 1951, 141; 1955, 141–3; & 1957, 15

Appendix Table 13-3.   Continued.

Tax Year	Tax Revenues[1]	Number of Returns[2]	Total Net Assets	Percent Stocks	Source
1949	$25.550M	2,281	$300,081M	Unknown	*TS* 1950, 157; 1951, 141; 1955, 141–3; & 1957, 15
1950	$29.920M	2,014	$297,587M	Unknown	*TS* 1950, 157; 1951, 141; 1955, 141–3; & 1957, 15
1951	$33.600M	2,044	$277.713M	Unknown	*TS* 1950, 157; 1951, 141; 1955, 141–3; & 1957, 15
1952	$38.208M	2,098	$326.179M	Unknown	*TS* 1950, 157; 1951, 141; 1955, 141–3; & 1957, 15
1953	$38.071M	2,304	$334.146M	Unknown	*TS* 1950, 157; 1951, 141; 1955, 141–3; & 1957, 15
1954	$39.138M	Not Available		Unknown	*TS* 1950, 157; 1951, 141; 1955, 141–3; & 1957, 15
1955	$44.768M	Not Available		Unknown	*TS* 1950, 157; 1951, 141; 1955, 141–3; & 1957, 15
1956	$66.607M	Not Available		Unknown	*TS* 1950, 157; 1951, 141; 1955, 141–3; & 1957, 15
1957	$79.709M	Not Available		Unknown	*TS* 1950, 157; 1951, 141; 1955, 141–3; & 1957, 15
1958	$71.608M	Not Available		Unknown	*TS* 1973, 202 & 1958, 15
1959	$72.535M	Not Available		Unknown	*TS* 1973, 202 & 1959, 19
1960	$88.431M	Not Available		Unknown	*TS* 1973, 202 & 1960, 17

Appendix Table 13-3. Continued.

Tax Year	Tax Revenues[1]	Number of Returns[2]	Total Net Assets	Percent Stocks	Source
1961	$84.879M	Not Available		Unknown	*TS* 1973, 202 & 1961, 17
1962	$84.580M	Not Available		Unknown	*TS* 1973, 202 & 1962, 17
1963	$87.143M	Not Available		Unknown	*TS* 1973, 202 & 1963, 19
1964	$90.671M[5]	Not Available		Unknown	*TS* 1973, 202 & *CYB* 1965, 981
1965	$88.626M[5]	Not Available		Unknown	*TS* 1973, 202 & *CYB* 1966, 990
1966	$108.352M[5]	5,673	$0.892480T	34.52%	*TS* 1973, 202; 1967, 124; & *CYB* 1967, 1046
1967	$101.106M[5]	6,184	$1.063671T	30.71%	*TS* 1973, 202; 1968, 162; & *CYB* 1968, 1035
1968	$102.192M[5]	6,688	$1.052908T	31.58%	*TS* 1973, 202; 1969, 174; & *CYB* 1969, 1066
1969	$112.377M	7,414	$1.193521T	30.73%	*TS* 1973, 202; *CYB* 1970–1, 1152 & *TS* 1970, 176
1970	$100.631M	5,893	$0.980984T	30.64%	*TS* 1973, 202; *CYB* 1970–1, 1152 & *TS* 1970, 176; & *TS* 1971, 200
1971	$119.835M	5,717	$1.018776T	33.91%	*TS* 1973, 202; 1972, 204; & *CYB* 1972, 1149
1972	$132.016M	5,980	$1.039893T	29.83%	*TS* 1973, 202 and 204; & *CYB* 1976–7, 971
1973	$71.594M[6]	Not Available		Unknown	*CYB* 1976–7, 971; *TS* 1974, 222
1974	$39.117M[6]	Not Available		Unknown	*CYB* 1976–7, 971; *TS* 1974, 222 & *TS* 1975, ???
1975	$24.701M[6]	Not Available		Unknown	*CYB* 1976–7, 971; *TS* 1974, 222 & *TS* 1976, ???

Appendix Table 13-3.   Continued.

---

[1] Succession duties related tax revenues for the 1942–1954 tax years were confirmed and are consistent with those presented in Perry (1955, 639).

[2] Non-resident or foreign estates were included, as follows: 1,315 (1945), 1,616 (1946), 2,290 (1947), 2,428 (1948), 2,217 (1949), 1,940 (1950), 2,042 (1951), 1,914 (1952), 1,938 (1953), . . . , 1,595 (1966), 1,563 (1967), 1,565 (1968), 1,609 (1969), 1,621 (1970), 1,801 (1971), and 1,944 (1972).

[3] The Dominion Succession Duty Act applied to estates of persons dying on or after June 14, 1941.

[4] Beginning January 1, 1948, all estates below $50,000 became exempt (explaining the significant decline in the number of estate tax returns).

[5] Includes provincial income taxes collected by the Canadian Taxation Division.

[6] Includes provincial succession duties and gift taxes collected by the Canadian Taxation Division.

Appendix Table 13-4. Descriptive Statistics from Canadian Gift Tax Returns.

Tax Year	Tax Revenues	Source:
1936	$0.194M	*DTC* 1943, 9
1937	$0.084M	*DTC* 1943, 9
1938	$0.374M	*DTC* 1943, 9
1939	$0.346M	*DTC* 1943, 9
1940	$0.398M	*DTC* 1943, 9
1941	$0.227M	*DTC* 1943, 9
1942	$0.264M	*DTC* 1943, 9
1943	$0.223M	*DTC* 1943, 9
1944	$1.547M	*DNR* 1944, 7
1945	$0.533M	*TS* 1946, 11
1946	$0.770M	*TS* 1947, 10
1947	$1.539M	*TS* 1947, 10
1948	$2.269M	*TS* 1948, 18
1949	$1.633M	*TS* 1949, 18
1950	$2.090M	*TS* 1950, 17
1951	$3.118M	*TS* 1951, 17
1952–64	Not Available	
1965	$7.365M[1]	*TS* 1967, 125

[1] Only 2,670 returns were filed for the 1965 tax year.

APPENDIX II

Supplemental Chapters

APPENDIX 2A
BASIC INDIVIDUAL FEDERAL
INCOME TAX RETURN FORMAT

INTRODUCTION

A basic discussion of the components of the U.S. individual federal income tax return is essential to the understanding of many of the historical, contemporary, and prospective issues addressed later. Pages 1 and 2 of the U.S. Form 1040 represent a summary sheet. Other U.S. individual federal income tax forms and schedules are used to generate detailed information, which is summarized on pages 1 and 2 of the Form 1040. The basic framework or summary, contained on pages 1 and 2 of the Form 1040, are represented in Fig. 2A-1 and are briefly explained below:

GROSS OR TOTAL INCOME

Gross income (also known as *total income*) includes wages and salaries, taxable interest and dividends, taxable state income tax refunds, alimony received, and business income (in the form of *net earnings* or *losses* from self-employment).

Also included in gross income are capital and other gains and losses, individual retirement account (IRA) and pension distributions, rental and royalty income and losses, farm income, unemployment compensation, social security benefits, and other income items.

DEDUCTIONS FROM/ADJUSTMENTS TO GROSS INCOME

Deductions from gross income are also referred to as *above-the-line* deductions or *adjustments* to income. The *line* refers to *adjusted gross income* (AGI).

233

	Gross or total income
less:	Above-the-line deductions or adjustments to gross income
equals:	Adjusted gross income (AGI)
less:	The larger of itemized or the standard deduction
less:	Personal exemption(s)
equals:	Taxable income (TI)
multiply by:	Tax rate or bracket (or select amount from tax table)
equals:	Tax liability
add:	Other taxes (including the SE tax)
less:	Tax credits (including FIT withheld or prepaid)
equals:	Tax due (or refund)

Fig. 2A-1. A Basic Individual U.S. Federal Income Tax (FIT) Return Format.

Deductions from/adjustments to gross income include deductions for: contributions to IRAs, student loan interest expenses, medical savings accounts (MSAs), moving expenses, 50% of the self-employment (SE) tax, the self-employed health insurance deduction, Keogh and self-employed SEP and SIMPLE plan deductions, penalties imposed by financial institutions on the early withdrawal of savings, and alimony paid.

Some of these deductions or adjustments are *phased out* (and eventually eliminated) for *high-income* taxpayers. These post-1990 phase-outs, summarized in Appendix Table 2-1, are shown for the 1991–2000 tax years. Phase-outs have the effect of increasing the affected taxpayer's marginal federal income tax rate.

ADJUSTED GROSS INCOME

Adjusted gross income (AGI) *equals* gross income (or receipts) *less* above-the-line deductions (or excluded items of income) from gross income. AGI is used to establish floors (and ceilings) for taxpayers itemizing their deductible personal expenses. AGI is also used, in its raw form or as a starting point, to develop modified AGI measures for determining phase-out ranges for other deductions (e.g. itemized deductions and personal exemptions) for *high-income* taxpayers.

Generally, the higher the taxpayer's AGI, the lower the potential tax savings generated from medical and dental expenses, casualty and theft losses, and miscellaneous itemized deductions.

ITEMIZED AND STANDARD DEDUCTIONS

Itemized deductions are personal expenses that U.S. tax law allows individual taxpayers to deduct in arriving at taxable income. AGI is reduced by itemized deductions.

Itemized deductions include medical expenses,[111] state and local income and property taxes, home mortgage interest,[112] charitable contributions, personal casualty losses,[113] and miscellaneous employee expenses.[114]

Taxpayers have the option of reducing taxable income (and federal income tax) by subtracting either their itemized or standard deduction. Taxpayers will, of course, select the *larger* of the two.

The standard deduction provides taxpayers with some minimum presumed amount of itemized deductions to:

(1) Simplify the computation of their tax liability,
(2) Eliminate lower income individuals from the tax rolls, and
(3) Ease the IRS administrative burden of auditing detailed itemized deduction for *all* taxpayers.

The standard deduction is based on taxpayer filing status and is comprised of a *basic* standard deduction plus any *additional* standard deduction (e.g. for blind or older taxpayers).[115] For each new tax year, the *basic* standard deduction amount is increased or inflation-indexed.

Generally, homeowners represent the largest group of individual taxpayers able to itemize. Renters are more likely to use the standard deduction, since home mortgage interest is frequently significant in amount and the largest single itemized deduction item. There are exceptions. For example, some taxpayers make very significant charitable contributions.

In order for a taxpayer to itemize, it must be profitable to do so. This is true only when itemized deductions exceed the standard deduction that cannot be taken if the taxpayer chooses not to itemize. Therefore, taxpayers itemize only when they have *excess* itemized deductions.

Medical expenses are deductible as an itemized deduction only for that portion of unreimbursed expenses exceeding 7.5% of the taxpayer's AGI. For example, a taxpayer with an AGI of $10,000 and $1,000 of unreimbursed medical expenses may use $250 ($10,000 *multiplied by* 7.5% *equals* $750; $1,000 *less* $750 *equals* $250) in determining whether or not to itemize deductions. In this case, the $750 is *excluded* from the taxpayer's itemized deductions. Therefore, the 7.5% of AGI represents a 7.5% of AGI *exclusion*. Similar *exclusions* exist for casualty and theft losses (10% of AGI plus $100 per incident) and miscellaneous deductions (2% of AGI), in arriving at itemized deductions.

Phase-out of Itemized Deductions

For *high-income* taxpayers, itemized deductions are subject to a phase-out when a taxpayer's AGI exceeds the inflation-indexed *applicable amount* (e.g. $126,600 for married taxpayers, filing jointly, for the 1999 tax year). For these taxpayers, itemized deductions otherwise allowable are reduced by the *lesser* of:

(1) 3% of the excess of AGI above the "applicable amount" or
(2) 80% of itemized deductions.

Therefore, it is possible for *high-income* taxpayers to lose up to 80% of deductible personal expenses.

PERSONAL EXEMPTION(S)

A *personal exemption* is a deduction provided for the taxpayer and the taxpayer's dependents. The *personal exemption* deduction is inflation-indexed and increases each year. This topic is discussed in detail in Chapter 6, where personal exemptions available over the entire history of U.S. federal income taxation are summarized in Appendix Table 6-1. The personal exemption amount for the 1999 tax year was $2,750. The personal exemption is subject to phase-out (and eventual elimination) for *high-income* taxpayers.

Phase-out of Personal Exemption(s)

For *high-income* taxpayers, the personal exemption is phased-out. The personal exemption is phased out for taxpayers, based on their filing status and adjusted gross income (AGI) levels (see Appendix Table 2-1). For example, the taxpayer's personal exemption deduction is reduced by 2% for each $2,500 (or portion of $2,500) that the AGI exceeds the threshold (e.g. for a single (SGL) taxpayer, $249,100 *less* $126,600 *equals* $122,500 *divided by* $2,500 *equals* 49 *plus* 1 *equals* 50 *multiplied by* 2% *equals* 100% phase-out). Note that this is *not* an extremely valuable deduction (e.g. if 2% *equals* $2,500, 2% *divided by* $2,500 *equals* 0.0008%).

Generally, these phase-outs (see Appendix Table 2-1) begin for taxpayers in federal income tax brackets of 31% or above (based on the assumption that AGI is approximately equivalent to taxable income).

TAXABLE INCOME

Taxable income (TI) is AGI *less* personal (non-business) deductions and exemptions (for dependents, age, and blindness). The U.S. tax system shifted from reliance on *net income* (1913–1943) to *taxable income* beginning with the 1954 tax year. This shift is discussed in Chapter 3.

FEDERAL INCOME TAX

The federal income tax (FIT) is imposed on TI. In the U.S., separate tax rates apply based on the taxpayer's *filing status* (e.g. single, married filing jointly, married filing separately, surviving spouse, and head of household).

In the U.S., the *unit of taxation* is the *household*. In Canada (see Appendix 2B) the *unit of taxation* is the *individual*.

OTHER TAXES

The self-employment (SE) tax is imposed in addition to the FIT. While the *employee* must pay both the employee's portion of FICA (7.65%) and FIT, the *self-employed* taxpayer must pay both SE tax (7.65% *multiplied by* 2 *equals* 15.3%) and the FIT.

For high-income taxpayers, the alternative minimum tax (AMT) may also be imposed in the calculation of FIT. The AMT prevents taxpayers from taking advantage of *tax preference items* (e.g. accelerated depreciation for a rental property) to achieve excessive tax savings, or to completely avoid the FIT by using aggressive tax planning strategies.

Additional taxes include those imposed on tip income, IRA or deferred compensation (including MSAs) and retirement plan distributions, advanced earned income credit (EIC) payments received from low-income taxpayers, and household employment taxes imposed on taxpayers employing others for domestic assistance.

TAX CREDITS

Having calculated the total tax liability, tax credits are applied against this liability. A tax credit is a dollar for dollar reduction in a taxpayer's FIT liability.

Tax credits include federal withholding taxes from employee salaries, estimated taxes paid by self-employed or other taxpayers, the EIC generated for low-income taxpayers, the child tax credit, amounts paid with extension requests, excess social security payments (or the alternative railroad retirement taxes),[116] and other payments made to the IRS.

TAX DUE/(REFUND AVAILABLE)

If the tax credits (payments and amounts withheld) are larger than the tax liability, a refund is due. If the tax credits do not completely offset the tax liability, a payment is due to the IRS.

APPENDIX 2B
CONTEMPORARY U.S./CANADIAN
COMPARISONS

It is not surprising that tax reform has followed rather similar patterns in the United States and in Canada. Geographic proximity and trade involvement alone prohibit extreme divergence, as do shared traditions of tax thinking . . . common patterns dominate, and future reforms will also be subject to similar trends (Musgrave & Wilson, 1992, 359).

. . . Canada has a national sales tax, while the United States does not; the United States has sharply higher social security taxes; Canada has much more revenue sharing and, generally, a more decentralized federal structure (Shoven & Whalley, 1992, ix).

INTRODUCTION

Historically, there have been many similarities between the U.S. (1913–) and Canadian (1917–) individual federal income tax systems. A significant contemporary departure occurred when the Canadian system was reformed in the late 1980s (i.e. 1988 Tax Reform).[117] Canadian *taxable* returns declined from 76.5% (13.1 million) for 1987 to 73% (12.8 million) for 1988 of all individual income tax returns filed for these tax years (*RC TS*, 1992, 69).

Canadian *personal exemption deductions* (through 1987) provided: (1) basic deductions for the taxpayer, (2) an additional deduction for age, (3) a spousal/ married or equivalent deduction, and (4) a deduction for dependents. These deductions were replaced by non-refundable tax *credits* (1988–) at the first tax rate/bracket of 17%. Beginning with the 1989 tax year, these tax credits are indexed annually by increases in the Canadian consumer price index (CPI) in excess of 3 percentage points. Therefore, the Canadian individual federal income tax system has discarded the progressivity feature of these deductions, providing for the same tax savings to all individual taxpayers, in absolute terms (*CTF*, 1990, 7:6).

Similarly, other individual federal income tax *deductions* (through 1987) were provided for (5) Canadian pension plan (CPP) or Quebec pension plan

(QPP) contributions, (6) unemployment insurance premium payments (paid by individual Canadian taxpayers), (7) pension income, (8) disability, (9) tuition fees, (10) education expenses, (11) eligible spousal transfers (the Canadian *unit of taxation* is the *individual*), (12) medical claims, and (13) charitable donations. These were also replaced, in form, by tax *credits* (1988–). However, the tax rates applied to both U.S. and Canadian taxpayers remain quite similar.

Figure 2B-1 graphically depicts the U.S. and Canadian individual federal income progressive tax rate structure. This illustration uses a Canadian taxable income range of $0–$100,000 (CAN) for the 1997 year as a basis for the U.S. taxable income comparison for the same tax year. Because U.S. tax rates are dependent on the taxpayer's filing status, the U.S. measures were presumed to be for the single U.S. taxpayer. This illustration also adjusts for approximate currency/purchasing power differences (e.g. $100 (CAN) ≈ $66.67 (U.S.)).

As Fig. 2B-1 suggests, the progressive Canadian tax rates, with their 17%, 26%, and 29% brackets (plus additional surtaxes of 3% on all federal tax payable and 5% on basic federal tax of $12,500 or more) are comparable to the first three tax brackets of their U.S. counterparts, of 15%, 28%, and 31%.[118]

There are, of course, many other differences between the U.S. and Canadian systems of individual federal income taxation (e.g. the Canadian harmonized

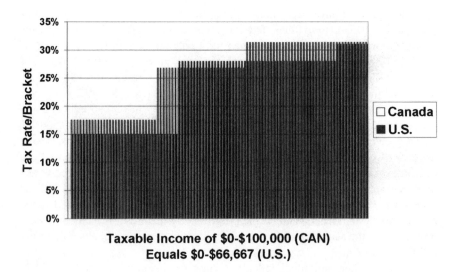

Fig. 2B-1. A U.S./Canadian Comparison for the 1997 Tax Year.

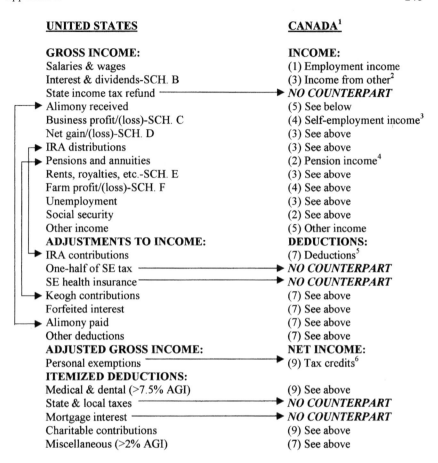

UNITED STATES	CANADA[1]
GROSS INCOME:	**INCOME:**
Salaries & wages	(1) Employment income
Interest & dividends-SCH. B	(3) Income from other[2]
State income tax refund ⟶	*NO COUNTERPART*
⟶ Alimony received	(5) See below
Business profit/(loss)-SCH. C	(4) Self-employment income[3]
Net gain/(loss)-SCH. D	(3) See above
⟶ IRA distributions	(3) See above
⟶ Pensions and annuities	(2) Pension income[4]
Rents, royalties, etc.-SCH. E	(3) See above
Farm profit/(loss)-SCH. F	(4) See above
Unemployment	(3) See above
Social security	(2) See above
Other income	(5) Other income
ADJUSTMENTS TO INCOME:	**DEDUCTIONS:**
⟶ IRA contributions	(7) Deductions[5]
One-half of SE tax ⟶	*NO COUNTERPART*
SE health insurance ⟶	*NO COUNTERPART*
⟶ Keogh contributions	(7) See above
Forfeited interest	(7) See above
⟶ Alimony paid	(7) See above
Other deductions	(7) See above
ADJUSTED GROSS INCOME:	**NET INCOME:**
Personal exemptions ⟶	(9) Tax credits[6]
ITEMIZED DEDUCTIONS:	
Medical & dental (>7.5% AGI)	(9) See above
State & local taxes ⟶	*NO COUNTERPART*
Mortgage interest ⟶	*NO COUNTERPART*
Charitable contributions	(9) See above
Miscellaneous (>2% AGI)	(7) See above

---

[1] As adapted from *Income Statistics* (Revenue Canada 1998, 166-185).

[2] Includes employment insurance benefits, interest and dividends, payments received from RRSPs, rental income, and taxable capital gains.

[3] Includes net business income from sole proprietors (including farming and fishing) and partnerships and workers' compensation.

[4] Includes Old Age Security and CPP or QPP.

[5] Includes RRSP contributions, union or professional dues, employee business expenses, and alimony paid.

[6] Includes medical expenses (>3 percent NI).

Fig. 2B-2. A Comparison of Contemporary U.S./Canadian Individual Tax Returns (1995).

sales tax (HST) is not included and the U.S. system of separate funding for Social Security and Medicare is not reflected).

Figure 2B-2 provides for a broad comparison of the U.S. and Canadian individual federal income tax return formats.

APPENDIX 3C
CHARITABLE CONTRIBUTIONS

In effect, the charitable deduction requires the treasury to *match* (emphasis added) private contributions ... Taxpayers who do not itemize their deductions were allowed to deduct some of their charitable contributions between 1982 and 1986, but this provision expired at the end of 1986 and was not extended by the 1986 act (Pechman, 1987, 94).

The large growth of private foundations, some of which have been suspected of abusing the tax exemption privilege, has led to concern about the charitable deduction (Pechman, 1987, 250).

INTRODUCTION

The charitable contributions deduction is not likely to become a topic of great interest in the near future. However, if the U.S. system in individual taxation were to be replaced by some of the flat tax or national sales tax proposals (discussed in Chapter 4), interest in this topic would increase. The elimination of the itemized deduction for charitable contributions might jeopardize the survival of many marginal tax-exempt organizations.

Charitable giving is viewed as a close substitute for government programs (Clotfelter & Steuerle, 1981, 403). Empirical investigations into the charitable deduction have included simulations of the alternatives of a charitable contribution: (1) tax credit or (2) "above-the-line" deduction (Clotfelter & Steuerle, 1981, 436). The Treasury Department recommended (1969 and 1984) that charitable contributions be allowed only for amounts above some AGI exclusion (e.g. for amounts in excess of 2% of AGI), but tax-exempt organizations were opposed to this change (Pechman, 1987, 98).

HISTORICAL BACKGROUND

Charitable contributions were first granted tax deductible status, as a *general deduction*, beginning with the 1917 tax year. The purpose of the deduction was to ensure that the (relatively recent imposition of the) individual income tax did not discourage charitable giving (Goode, 1976). Individual taxpayers were permitted to deduct a maximum of 15% of *net income*.[119]

The *Individual Income Tax Act of 1944* retained the 15% charitable contributions limitation, but beginning with the 1944 tax year, this and similar limitations were applied to the individual taxpayer's *adjusted gross income* (AGI) instead of the taxpayer's *net income*.[120] Figure 3-1 provided an illustration of this 1943/1944 shift from total income to AGI.

Beginning with the 1952 tax year,[121] the allowable deductions for charitable contributions were increased from 15% to 20% of AGI, for taxable years beginning on or after January 1, 1952 (IRS SOI, 1952, 6). The IRC54 allowed a special *additional* deduction of up to 10% of AGI for contributions made to churches and tax-exempt organizations. Therefore, the total charitable contribution deduction, if restricted to tax-exempt organizations, could reach 30% (i.e. the original 20% *plus* the additional 10%) of AGI.[122] The *Revenue Act of 1964* provided for the carryover of excess charitable contributions to subsequent years.

TRA69 provided for a post-1969 tax year increase, to a maximum of 50% of AGI, of the ceiling on the individual charitable contributions deduction.[123] Post-TRA69, contributions made to private non-operating foundations were subject to a 20% limitation; contributions of certain appreciated and unrealized capital gain property were subject to a 30% limitation; and taxpayer contributions exceeding both 50% and 30% limitations were provided with a five-year carryover (IRS SOI, 1973, 183).

Under contemporary law, individual taxpayers able to itemize may continue to deduct varying categories of charitable contributions up to 50%, 30%, or 20% of AGI. However, for the post-1989 tax years, these (and other) itemized deductions are subject to reduction or phase-out (see Chapter 2) if the taxpayer's AGI exceeds specified amounts.

DESCRIPTIVE STATISTICS OF CHARITABLE CONTRIBUTIONS

Appendix Table 3C-1 provides descriptive statistics on aggregate charitable contributions for the 1917–1996 tax years, both in dollar amounts and as a percentage of net income or adjusted gross income.

Appendix Table 3C-2, using the descriptive statistics for the 66 tax years contained in Appendix Table 3C-1, provides the results of a Mann-Whitney U test for the comparability of mean charitable contributions under 15% (1917–1951), 20% (1952 and 1953), 30% (1954–1969), and 50% (1970–1996) ceilings (based on NI or AGI, as appropriate) established over the 1917–1996 tax years.

Mean charitable contribution percentages were significantly different (at the 10% level) between the 20% and 30% periods and the 50% and all prior periods. However, as Appendix Table 3C-2 results suggest, the increased deductibility percentage of charitable contributions, in and of itself, did not result in increased mean contributions (1.9% to 1.5% for the 1951 to 1952 change from 15% to 20% deductibility and 2.0% to 1.8% for the 1969 to 1970 change from 30% to 50% deductibility).

SUMMARY

Charitable contributions first became deductible for the 1917 tax year at 15% of *net income*. Beginning with the 1944 tax year, this 15% ceiling was applied to AGI. Increased to a ceiling of 20% of AGI in 1952, an additional 10% (for a total of 30%) of AGI could be deducted if charitable contributions were made to churches or tax-exempt organizations. Beginning with the 1970 tax year, individual taxpayers could deduct certain classes of charitable contributions of up to 50% of AGI. Presently, varying classes of charitable contributions remain deductible, as an itemized deduction, at 50, 30, and 20% of AGI ceilings. For the post-1989 tax years, itemized deductions, which include charitable contributions, are subject to phase-out for high-income taxpayers.

APPENDIX 3D
MEDICAL AND DENTAL EXPENSES

... medical deductions ... are heaviest in the lowest income classes (Pechman, 1987, 92).

INTRODUCTION

The medical and dental expenses deduction has not been a controversial topic in recent years, and is not likely to become one in the near future. However, if the U.S. system in individual taxation were to be replaced by some of the flat tax or national sales tax proposals (discussed in Chapter 4), interest in the deductibility of medical and dental expenses would peak. Alternatively, if the U.S. were to adopt a system of socialized medicine, such as that used in Canada, medical and dental expenses would simply disappear, along with the *directly traceable* cost to the taxpayer.

Large, unusual, and necessary medical and dental expenses affect a family's *ability to pay* tax. These expenses are often involuntary, unpredictable and may exhaust a large proportion of a taxpayer's total income for a particular year. It is for this reason that taxpayers are permitted to deduct uninsured medical (and dental) expenses (Pechman, 1987, 92). The contemporary 7.5% exclusion (see Appendix 3D) of medical expenses represents an arbitrary dividing line between "usual" and "extraordinary" medical expenses (Pechman, 1987, 98), where the intent is to eliminate the tax subsidy associated with the former, but not the latter.

HISTORICAL BACKGROUND

The *Revenue Act of 1942* provided for the initial deductibility of unreimbursed medical and dental expenses, beginning with the 1942 tax year. The deduction was limited to medical and dental expenses in excess of a 5% *exclusion* of these expenses (see Appendix 3D), based on the taxpayer's *net income*. In addition, *ceilings* of $1,250 (if one exemption was claimed) or $2,500 (if two or more exemptions were claimed) were established (IRS SOI, 1942, 23).

The *Individual Income Tax Act of 1944* retained the 5% floor, but beginning with the 1944 tax year, the limitation was applied to the individual taxpayer's *adjusted gross income* (AGI) instead of the taxpayer's *net income*.[124]

The *Revenue Act of 1948* increased the ceilings in the case of a joint return for a husband and wife, where the maximum was established at $5,000 (IRS SOI, 1948, 8). The *Revenue Act of 1951* liberalized the medical deduction for taxpayers age 65 or older, by removing the 5% floor. It permitted these taxpayers to deduct their entire medical costs (IRS SOI, 1952, 8). The IRC54: (1) reduced the floor of deductible medical expenses (from 5%) to 3% of AGI; (2) continued to provide for the inclusion of expenses for drugs and medicine, but only to the extent that they exceed 1% of AGI (applicable to all taxpayers, including those 65 years of age or older); and (3) raised the limitations or ceilings on deductions to $2,500 per exemption (other than the exemptions for age and blindness) and $5,000 per return or $10,000 on a joint return, a return for a head of household, or a return for a surviving spouse (IRS SOI, 1954, 17).

The *Technical Amendments Act of 1958* further liberalized the medical deduction for disabled persons aged 65 or over. Generally, a medical expense ceiling of $15,000 was in effect for all taxpayers. However, this maximum was doubled to $30,000 for married taxpayers, where both were age 65 or over and disabled (IRS SOI, 1958, 21–22).

Beginning with the 1960 tax year, the deduction of medical expenses for a dependent parent age 65 or over became fully deductible. A new Form 2948 became available to assist with these dependent parent and other medical and dental expense computations (IRS SOI, 1960, 124–125).

PL 87-863, effective for tax years after 1961, revised the ceilings for medical and dental expense deductions. Generally, ceilings were doubled.[125]

Beginning with the 1967 tax year, one-half (up to $150 per return) of medical insurance premiums were deductible without consideration of the 3% of AGI exclusion. The 1% and 3% of AGI exclusions for medicines/drugs and other medical expenses became applicable to taxpayers age 65 or older, and the ceilings on the maximum medical expense deduction were eliminated (IRS SOI, 1967, v; CCH, 1967, 3).

The ERTA81 provided for additional changes. For the 1982 tax year, individual taxpayers could deduct the greater of: (1) medical and dental expenses in excess of 3% of AGI, or (2) one-half of insurance premiums (up to a $150 ceiling). Beginning with the 1983 tax year, a taxpayer could only deduct combined medical and dental expenses and insurance premiums in excess of 5% of AGI (IRS SOI, 1983, 2).

TRA86, beginning with the 1987 tax year, increased the exclusion applied to medical and dental expenses to 7.5% (from 5%) of AGI (IRS SOI, 1987, 4).

Also beginning with 1987, certain self-employed taxpayers were permitted to deduct as an *adjustment to income* (or *above-the-line* deduction) up to 25% of their medical insurance payments (IRS SOI, 1987, 7).[126] As a result of the various tax acts, these percentages were increased to 30 (1996), 40 (1997), 45 (1998), 60 (1999–2001), 70 (2002) and 80% (2003–) of the total cost of health insurance for self-employed taxpayers.

SUMMARY

Medical and dental expenses were not deductible as an itemized deduction until the 1942 tax year. The first 5% of the taxpayer's *net income* was excluded (floor) and these deductions were limited (ceiling) to $1,250 for one exemption or $2,500 for two or more exemptions. The floor was applied to taxpayers' AGI, beginning with the 1944 tax year. Beginning with the 1954 tax year, the floor was reduced to 3% of AGI. A separate floor or exclusion of 1% of AGI was implemented for medicine and drugs. Beginning with the 1967 tax year, limited amounts of medical insurance premiums became deductible, and the ceilings for the medical expense deduction were eliminated. For the 1983 and future tax years, the exclusion increased to 5% of AGI and, beginning with the 1987 tax year it was increased still further, to 7.5% of AGI.

APPENDIX 3E
INCOME-AVERAGING

The use of an annual accounting period, combined with progressive income taxes, results in a heavier tax burden on fluctuating incomes than on an equal amount of income distributed evenly over the years (Pechman, 1987, 127).

INTRODUCTION

Historically, income-averaging has been used to minimize or eliminate the impact of *progressive* tax system-based inequities that affect taxpayers with fluctuating income. However, as the contemporary U.S. system of individual taxation has become less progressive, the need for income-averaging has been minimized or eliminated. Therefore, unless (or until) our current tax rate structure is replaced with a more progressive one, income-averaging is not likely to return as a topic for controversy.

HISTORICAL BACKGROUND

The *Revenue Act of 1964* introduced income-averaging (effective for the 1964–1986 tax years). Income-averaging (computed on the Form 1040, Schedule G) was first available for the 1964 tax year. It was designed to spread income over a five-year period and benefit individuals with a unusually large amounts (or high fluctuations) of income for any one taxable year. The 1964 implementation of income-averaging coincided with a significant reduction in marginal tax rates (i.e. from 20%–91% for the 10 years preceding 1964, to 16%–77% for the post-1963 tax years) (IRS SOI, 1964, 65).

All types of taxable income qualified for income-averaging, except income from gifts and inherited property, wagering income, and net long-term capital gains, but the largest number of returns using income-averaging were those of taxpayers whose principal source of income was salaries and wages. However, the largest amounts of tax savings were realized by taxpayers whose principal source of income was from a business or profession (e.g. sole proprietors) (IRS

251

SOI, 1965, 52). Generally, tax savings, as a% of income tax before income-averaging, were higher at the lower income levels, but this was principally due to the progressive nature of the individual federal income tax rates and the tendency for higher incomes to reflect capital gains as a major source of income (IRS SOI, 1969, 121).

Beginning with the 1970 tax year, taxpayers using the alternative tax computation for capital gains were prohibited from also using income-averaging. However, the income-averaging rules were liberalized for the post-1969 tax period, providing for the averaging of income items previously excluded (i.e. long-term capital gains, wagering, and gift income). For the post-1970 period, taxpayers choosing the income-averaging computation were also prohibited from using the special rates applying to "earned" (later referred to as "personal service") income under the maximum tax computation (see MAXTAX discussion in Appendix 3F) (IRS SOI, 1971, 123). The net effect of these changes resulted in a significant increase in the number of returns using income-averaging (from approximately 600,000 for the 1969 tax year to approximately 1 million for 1970) and a significant decline in the number of returns using the alternative tax computation (from more than 100,000 in 1969 to less than 70,000 for the 1970 tax year) (IRS SOI, 1970, 149–150).

SUMMARY

The number of returns using income-averaging increased consistently from 1964–1981. The tax savings achieved through income-averaging followed a similar pattern, also peaking for the 1981 tax year. ERTA81 (see Appendix 3D) reduced maximum marginal federal income tax rates from 70% to 50%. This and future tax rate/bracket reductions, combined with increasingly restrictive qualifying requirements for the use of income-averaging, contributed to the obsolescence and reduced importance of income-averaging. Income-averaging was repealed for the 1987 tax year.

APPENDIX 3F
THE MAXIMUM TAX

INTRODUCTION

The maximum tax, like income-averaging, was been used to minimize or eliminate the impact of the *progressive* tax system-based inequities, but on *earned* income. However, as the contemporary U.S. system of individual taxation has become less progressive, the need for the maximum tax ceiling was eliminated. Therefore, unless (or until) our current tax rate structure is replaced with a more progressive one, a similar need is not likely to arise or become a topic for controversy.

HISTORICAL BACKGROUND

TRA69, effective for the 1971–1981 tax years, contained a provision to ease the tax burden on taxpayers generating high-income largely from *earned net income* (e.g. salaries and wages). The *maximum tax* (MAXTAX) applied a maximum tax rate of 60% (instead of the top rate of 70%) to income from earnings (IRS SOI, 1971, 123).

This MAXTAX rate was decreased to 50% for the 1972 tax year. The tax savings associated with the 1972 tax year approximated $271 million (IRS SOI, 1972, 145). The number of taxpayers taking advantage of this special tax rate increased from 9,185 in 1971 to 88,085 in 1972 (IRS SOI, 1971, 130; 1972, 145). The TRA76 expanded the impact of the MAXTAX by modifying the computation to include all *personal service net income* (1977–). The number of taxpayers electing to use the MAXTAX increased to a peak of 483,293 (the all-time high of 0.51% of all individual tax returns filed) for its final tax year of 1981 (IRS SOI, 1981, 72). The MAXTAX was not available to married taxpayers filing separate returns or to taxpayers electing the use of income-averaging (applicable for the 1964–1986 tax years).

253

SUMMARY

The ERTA81 resulted in the repeal of the MAXTAX on *personal service* income. Beginning with the 1982 tax year, when maximum regular marginal tax rates for individual taxpayers were reduced to the (previously available) MAXTAX rate of 50%, the MAXTAX became obsolete.

APPENDIX 3G
INTEREST DEDUCTIONS

Deductions for interest are justifiable when the interest is paid on a loan used to produce taxable income. A substantial proportion of the interest deducted on tax returns, however, has been for loans on homes . . . people with high incomes have often deducted interest on debt incurred to carry investment assets that produce no current income, causing a mismatching of income and expenses (Pechman, 1987, 94–5).

INTRODUCTION

Existing laws regarding the deductibility of various classifications of interest expense appear to be quite stable. Unless the present individual Federal income tax system is replaced with a non-itemizing flat tax proposal or a national sales tax (discussed in Chapter 4), the underlying rationale for the current classification scheme and status of interest expense deductibility is not likely to vary greatly in the near future. This is particularly the case with the home mortgage interest deduction, the elimination of which has been only briefly discussed in recent decades . . . followed by speedy political retreats as this periodic "testing of the waters" draws very rapid fire.

The 1986 tax reform act phased-out deductions for interest on consumer debt and limited deductions for investment interest (i.e. loans to purchase securities), but retained almost complete deductibility of home mortgage interest and investment interest (Pechman, 1987, 95).

HISTORICAL BACKGROUND

The five categories of interest include: (1) business interest (fully deductible and without limitations), (2) passive activity interest (e.g. rental property interest expense), (3) investment interest (e.g. interest expense associated with the financing of investments), (4) personal interest (e.g. credit card, car loans and personal loans), and (5) qualified residence interest (limited for high-income taxpayers, beginning with the 1987–88 tax years).[127]

Business interest has always been fully deductible. This is not likely to change. Passive activity interest represents a component of passive activity losses (PALs), a topic separately covered in Appendix 3I. The other three categories of interest are discussed below.

Investment Interest

Investment interest is that interest paid or accrued on debt used to purchase or finance property held for investment. TRA69 changed the treatment of *investment interest*. Previously, investment interest was deductible, generally without limitation, as an itemized deduction. Presently, the deductibility of investment interest may be limited. These limitations are intended to prevent high-income taxpayers from aggressively pursuing the deferral of tax through a *mismatching* of revenues and expenses. It limits the economic incentives associated with the generation of large and immediately tax deductible interest deductions on the *current* financing costs associated with *long-term* investments.

The excess of investment interest over net investment income became subject to a minimum tax for the 1970 and 1971 tax years. Beginning with the 1972 tax year, it became subject to new deduction limitations (and was reported on a new Form 4952). The deductibility of investment interest was limited to: (1) $25,000 ($12,500 for married taxpayers filing separately), *plus* (2) net investment income (i.e. income from dividends, rents, and interest, *less* allowable expenses associated with the production of these income items), *plus* (3) the excess of net long-term capital gains over net short-term capital losses, *plus* (4) the excess of rental expenses over rental income attributable to property subject to a net lease, *plus* (5) one-half of the amount by which investment interest exceeded the sum of items (1), (2), (3), and (4). Investment interest disallowed for the current tax year could be carried over to the following tax year, subject to varying limitations and carryforward reductions (IRS SOI, 1973, 186).

Beginning with the 1993 tax year, the net investment income used to increase the deductible portion of investment interest could not include net capital gain, taxed at the 28% maximum capital gains tax rate (IRS SOI, 1993, 112).

Personal Interest

TRA86 provided for the phase-out of the deductibility of *personal interest*. Prior to the 1986 tax year, personal interest was fully deductible. The deductibility of personal interest was reduced to 65% (1987 tax year), 40%

(1988 tax year), 20% (1989 tax year), 10% (1990 tax year), and completely eliminated (1991 and future tax years) as an itemized deduction. A very similar phase-out occurred for passive activity losses (PALs) (see Appendix 3I).

The phase-out and elimination of the personal interest deduction came at a time when concerns were high regarding both: (1) deficit spending and the growth in the U.S. national debt, and (2) the growth in consumer debt. The former was difficult to control; the latter was easier to control through the manipulation of the after-tax cost of consumer financing. Concerns during this period also included the competition for available funds between the public and private sectors, and a possible rise in the cost of government borrowing.

Mortgage Interest

TRA86 provided for a limitation on the deductibility of home mortgage interest. Taxpayers with a mortgage *before* October 14, 1987 could deduct all interest. Taxpayers who got a mortgage *after* October 13, 1987 could effectively deduct all interest for mortgages up to a ceiling of $1.1 million (IRS SOI, 1988, 5).

The limitation on the deductibility of home mortgage interest expense, through the $1.1 million ceiling (1987–), preserves the politically (and economically) desirable association of home ownership and related tax benefits/savings for lower- and middle-class taxpayers, while preventing excessive tax savings for high-income taxpayers. In 1992, nearly 27 million individual tax returns claimed $197 billion in mortgage interest deductions (Pearlman, 1996, 422).

All do not agree with the importance of the desirability of home ownership as subsidized through the mortgage interest deduction. Some economists have suggested that a desirable solution to the differing tax benefits afforded home owners and renters be eliminated through: (1) the complete elimination of the home mortgage interest deduction or, (2) the institution of a comparable tax benefit based on the capitalization of rent equivalents. However, the elimination of the home mortgage interest deduction is not likely to be politically feasible in the near future.

SUMMARY

The limitation of investment interest and the ceiling on home mortgage interest evolved as equity issues. Relatively high-income taxpayers are prevented from benefiting disproportionately from these itemized tax deductions. Alternatively, the elimination of the personal interest deduction was a function of concerns over increasing debt levels in both public and private sectors.

APPENDIX 3H
THE TWO-EARNER MARRIED COUPLE DEDUCTION

> ... in 1981 Congress provided a special deduction for two-earner couples ... When the number of tax brackets and tax rates was reduced in 1986, Congress decided to eliminate the two-earner deduction. Although lower tax rates moderate the disadvantages of two-earner couples, they did not eliminate it entirely. Thus, on equity grounds, the two-earner deduction should be restored (Pechman, 1987, 107).

INTRODUCTION

The two-earner married couple deduction was intended to minimize what has come to be referred to as the "marriage tax penalty" (MTP). Most tax researchers and policymakers link the MTP to the separate progressive tax rate schedules for single and married taxpayers, beginning with the 1971 tax year (Cataldo, 1996, 611).

This issue is a complex one. It has resurfaced frequently in the past two decades. Unless resolved, the MTP will continue to remain an emotional political issue. For this reason, the MTP is addressed as a separate topic in Chapter 8.

HISTORICAL BACKGROUND

ERTA81 provided for tax laws to be phased-in over several years. Included in ERTA81 was an *above-the-line* deduction (see Appendix Fig. 2-1) to reduce the effects of the marriage tax penalty (MTP).

Beginning with the 1982 tax year, working married couples were provided an above-the-line deduction to gross income of 5% of the lesser of: (1) $30,000, or (2) *earned income* of the spouse with the lesser earnings (IRS SOI, 1982, 12). Reported on the Schedule W, this adjustment to gross income was intended to reduce the MTP associated with two-earner households. For the 1982 tax year, more than 21 million tax returns (almost 23%) of the tax returns filed, took advantage of this adjustment.

Beginning with the 1983 tax year, the percentage deduction/adjustment was increased from 5% to 10%. For the 1983 tax year, more than 23% of the tax returns filed include a Schedule W and an above-the-line deduction/adjustment for two-earner households (IRS SOI, 1983, 12 and 22). Participation increased to more than 24% (and approached 25%) for the 1984, 1985, and 1986 tax years (IRS SOI, 1984, 18 and 28; 1985, 20 and 29; and 1986, 24 and 34). This above-the-line deduction was not available for 1987 and future tax years.

SUMMARY

The two-earner married couple deduction (and the Schedule W) was repealed for the 1987 and future tax years. Though discussed frequently by politicians, the marriage tax penalty has not seen any similar direct relief efforts since the abolition of the Schedule W and the above-the-line deduction of 1982–1986.

APPENDIX 3I
PASSIVE ACTIVITY LOSSES

Beginning with 1987, losses generated by any trade, business, or rental activity in which the taxpayer did not "materially participate" were considered "passive" losses and could be used to offset income from other passive activities only. Under the previous law, any business losses could be used to offset income from any other activity (IRS SOI, 1987, 6).

In the Tax Reform Act of 1986, Congress tried to end most tax shelters. Top-bracket taxpayers got a significant tax rate cut, down from 50% in 1985 and down from 70% in 1981. In exchange, however, they were to give up tax shelters (Johnson, 1996, 380).

INTRODUCTION

The early 1980s saw tremendous growth in the formation and marketing of limited partnership interests. Frequently, these *investments* were designed to provide immediate tax savings, taking advantage of the individual tax return flow-through features of investment tax credits (ITCs) or the short depreciable lives made available through the accelerated cost recovery system (ACRS). Many of these investments were appropriately characterized as "abusive tax shelters" because they produced profits, but only in the form of tax deferral/savings.

Johnson (1996, 381) notes that the "primary siege gun against shelters in 1986 was the passive activities limitations of section 469" of the Internal Revenue Code. Taxpayers entering into passive activities prior to October 23, 1986 were provided with a five-year phase-in of passive activity loss (PAL) limitations, permitting a declining percentage of passive losses for use as an offset for nonpassive income and taxes. The phase-out resulted in a decrease from 100% (1986) to 0% (1991) for PALs (i.e. 100% for 1986 and prior tax years, 65% for 1987, 40% for 1988, 20% for 1989, 10% for 1990, and 0% for 1991 and future tax years).

The passive activity loss (PAL) rules were designed, in large part, to reduce or eliminate the tax deferral available to high-income taxpayers. Therefore, PAL rules arose as an equity issue, to prevent high-income taxpayers from escaping taxation.

HISTORICAL BACKGROUND

TRA86 imposed new rules affecting the deductibility of PALs.[128] Beginning with the 1987 tax year, losses generated from trade, business, or rental activity in which the taxpayer did not "materially participate" were considered "passive" losses and were restricted for use as an offset against the income from other passive activities.

Exceptions were made for rental real estate activities. If actively involved, taxpayers could deduct net losses of up to $25,000 from nonpassive income (e.g. wages or dividends), depending on the taxpayer's AGI (IRS SOI, 1987, 6).

Taxpayers who entered into passive activities before October 23, 1986 were provided with a five-year phase-in for the loss limitation. This phase-out of the deductibility of passive losses against nonpassive income was very similar to that provided for personal interest (see Appendix 3E).

The deductibility of PALs against "nonpassive income" was reduced to 65% (1987 tax year), 40% (1988 tax year), 20% (1989 tax year), 10% (1990 tax year), and completely eliminated (1991 and future tax years) as an offset against nonpassive income items. A very similar phase-out occurred for personal interest (see Appendix 3E).

Disallowed or non-deductible PALs could, in many cases, be carried forward. Most of the PALs were concentrated in partnerships and rental activities on the individual taxpayer's Schedule E (IRS SOI, 1988, 5).

A PRELIMINARY STATISTICAL ANALYSIS

Appendix Table 3I-1 (N = 17) contains summary descriptive statistics relating to the period preceding, including, and following (1980–1996) the active phase-out (1987–1990) of PALs. Rental profits/(losses) consistently amounted to less than 1% of aggregate individual FIT taxpayer AGI over the entire period (e.g. 0.9% for the 1985 tax year).

Figure 3I-1 clearly illustrates the impact of (1) the liberalization of the depreciable lives of assets, including rental properties, on rental profits/(losses) and (2) the reversal of this policy, combined with the imposition of PALs phase-out legislation (TRA86). Rental losses increased through the 1986 tax year, then declined, as PALs were phased-out. Rental profits have been generated for the 1993–1996 tax years.

U.S. Rental Profits/(Losses)

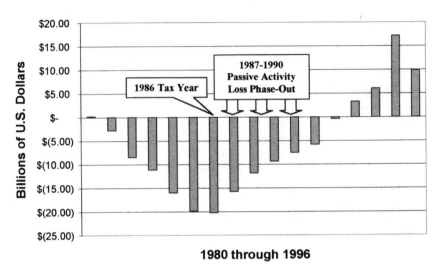

Fig. 3I-1. U.S. Retail Profits/(Losses).

Appendix Table 3I-2 contains the results of some exploratory statistical analyses. Our research question:

Did the PAL phase-out reduce (increase) taxable rental profits (losses), as Fig. 3I-1 suggests?

First, a preliminary correlation matrix was developed for the measures presented in Appendix Table 3I-1, where relations between PALs Allowed (PCTALLOW), Tax Returns with Rents (RENTRTRN), Percent Affected (PCTRENT), Rental Amounts (RENTAMT), and Pct AGI (PCTAGI) variables suggested high correlation and the inclusion of the PCTAGI[129] and PCTAL-LOW variables. The relation for the former was direct and stronger ($r = 0.972$). The relation for the latter was inverse and weaker ($r = -0.65$). (The RENTRTRN and PCTRENT measures also produced a strong correlation ($r = 0.97$), but were nearly equivalent measures).

Next, the regression model presented in Equation (3I-1) was developed, as follows:

$$RENTAMT = Intercept + PCTAGI + PCTALLOW + Error \qquad (3I\text{-}1)$$

Application of the C_p-statistic was employed for the 3 possible variations of the regression model presented in Eq. (3I-1), and suggested that the full model

(model 3) provided for the "best fit". These results are also presented in Appendix Table 3I-2.

Rental amounts as a percentage of AGI, at amounts never achieving even 1% of aggregate AGI, provided the greatest explanatory power when regressed against the RENTAMT dependent variable. Though model 2 (see Appendix Table 3I-2) produced the hypothesized and inverse relation between the RENTAMT and PCTALLOW variables (adj-R^2 = 38.8% in Appendix Table 3I-2), when used as the only independent variable, the model 3 results generated a reversal of this inverse relation, which also remained significant (p = 0.070 in Appendix Table 3I-2).

These counterintuitive results suggest that the regression model (model 3) presented in Eq. (3I-1) is under-/mis-specified. Though our purpose is not one of model-building. Statistical analyses are not necessary to see the strong relation between the phase-out of PALs and rental profits/(losses). We have achieved this objective in presenting Fig. 3I-1. Clearly, the use of a transform (e.g. the logarithm of AGI) or the exclusion of rents (endogenous) from AGI are more common in academic statistical model building efforts. Our purpose it to suggest that further inquiry into those components of taxpayer behavior, perhaps requiring the further decomposition of AGI, would provide insights into what alternative measures individual taxpayers adversely affected by TRA86 took in anticipation of the phase-out of PALs. The result would provide insights into individual taxpayer behaviors toward future tax policy changes affecting, primarily, high-income taxpayers. The descriptive statistics contained in Appendix Table 3I-1 provide a starting point.

SUMMARY

The deductibility of PALs, like personal interest, was phased-out beginning with the 1987 tax year. However, unlike personal interest, which is no longer deductible (1991–), PALs remain deductible against *passive income*. Some additional discussion of PALs (and descriptive statistics) is provided in Chapter 9.

APPENDIX 4J
THE MOVE FROM A *CLASS* TAX TO A *MASS* TAX

Between 1929 and the middle 1950s, says Seltzer, "the Federal individual income tax was transformed from one that applied to only 4% of the population, including taxpayers and their dependents, and to only 30% of individual incomes, to one that covered about 70% of the population and more than four-fifths of the greatly enlarged total of individual incomes" (Groves, 1963, 19, citing Seltzer, 1959).

It is clear enough that in the interlude between 1929 and the middle 1950s the income tax evolved from a *class* (emphasis added) to a *mass* (emphasis added) tax (Groves, 1963, 20).

"The Current Tax Payment Act of 1943" included the only full-fledged federal amnesty on personal income taxes to take place this century, granting taxpayers forgiveness of 75% of the lower of a taxpayer's 1942 or 1943 liabilities. "Pay as you go" became the rule nationwide in July of 1943 . . . (Shlaes, 1999, 7).

INTRODUCTION

Historically, taxes have been increased or imposed during periods of war. This appendix provides a broad overview of the historical trends or patterns in marginal individual federal income tax rates. For a summary of the entire history of U.S. minimum and maximum individual marginal federal income tax rates (1913–1993) see Cataldo (1994, 169–170).

The first year of the contemporary U.S. individual income tax was 1913. The lowest tax rate or bracket, for those individuals required to file a federal income tax return, was 1%. However, the maximum combined normal tax and surtax rates could reach a combined marginal tax rate of 7%.

The need to finance the cost of World War I led to marginal tax rate ceilings in excess of 60% for the 1917–1921 tax years. These ceilings declined through the mid-1920s, but increased sharply during the post-1929 stock market crash and depression era.

RISING WWII TAX RATES – FROM A *CLASS* TAX TO A *MASS* TAX

In the U.S., the two separate *Revenue Acts of 1940*, the *Revenue Act of 1941*, and the *Revenue Act of 1942*: (1) reduced personal exemptions and imposed surtaxes and/or increased marginal rates, and (2) imposed a steeply graduated excess profits tax, which targeted profits resulting from the defense buildup. By 1944 the maximum marginal individual federal income tax rate or bracket had risen to 94%. Approximately 74% of the U.S. population was required to file a federal income tax return (Pollack, 1996, 64). The growth in that portion of the population affected by the income tax has been referred to as a move from a *class tax* to a *mass tax* (Groves, 1963, 20).

Figures 4J-1 and 4J-2 graphically depict the growth in income tax returns filed by individuals for the U.S. (1913–1996) and Canada (1921–1995),

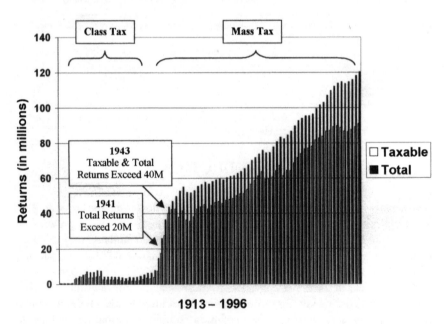

Fig. 4J-1. U.S. Taxable and Total Individual Federal Income Tax Returns (1913–1996).

Notes: For tax returns with net income of $3,000 or more (1913–1916), $1,000 or more (1917–1936), and excluding fiduciary/estate and trust returns (Forms 1041) for the tax years following 1936. M = Million.

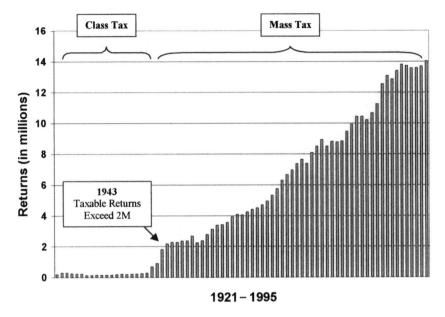

Fig. 4J-2. Canadian Taxable Individual Federal Income Tax Returns (1921–1995).
M = Million.

respectively. Both countries were involved in WWII and these graphical representations reflect the needs of both countries to finance their involvement in the war effort.

For both the U.S. and Canada, the number of individual income tax returns remained stable or flat through the 1930s. A sharp increase in individual taxpayer participation for both countries occurred in the early 1940s.

THE POST-WWII PERIOD

A reversal of the growth in U.S. marginal federal income tax rate ceilings did not occur in earnest until the early 1980s. The maximum individual marginal federal income tax rate remained above 90% for the 1951–1963 tax years. For the 1964–1981 tax years, statutory ceilings declined from 77% to 70%.

Further reductions led to a maximum marginal tax rate of 50% (1982–1986), to a "blended" or "transitional" rate of 38.5% (1987), followed by a ceiling of 33% (1988–1990), and finally to a recent historical low of 31% (1991 and

1992). For the 1993 and contemporary tax years the maximum statutory individual federal income tax rate in the U.S. has remained stable at 39.6%.

SUMMARY

The U.S. individual income tax has evolved from a tax system intended to apply only to the very wealthy (*class* tax) to one applying to the majority of the U.S. population (*mass* tax). Initially, increased participation was used to finance wars (e.g. WWI, WWII, Korea, Vietnam), but the U.S. individual income tax has evolved to a system used to finance social programs and to promote horizontal and vertical equity among the U.S. population.

   In recent decades, individual federal income tax rates have declined from a ceiling of 77% to 39.6%. Flat tax rate proposals might reduce ceilings even further, but raise equity issues. They fail to maintain the progressive features still present in the U.S. system of individual income taxation.

APPENDIX 4K
TAXATION OF THE INTERNET

INTRODUCTION

The Internet Tax Freedom Act (ITFA[130]) in its original form[131] was introduced by Representative Christopher Cox (R-CA) in early 1997. A compromised version (announced on March 19, 1998) was signed into law on October 21, 1998. It placed a three-year moratorium on the taxation of e-commerce, as follows:

> No State or political subdivision thereof shall impose any of the following taxes during the period beginning on October 1, 1998, and ending 3 years after the date of the enactment of this Act:
>
> (1) taxes on Internet access, unless such tax was generally imposed and actually enforced prior to October 1, 1998; and
> (2) multiple or discriminatory taxes on electronic commerce.

THE ADVISORY COMMISSION ON ELECTRONIC COMMERCE

On October 15, 1999, the 19 Congressionally appointed members of the Advisory Commission on Electronic Commerce (ACEC) issued an invitation to the public to submit proposals for consideration, presentation, and/or inclusion in the ACEC's "Policies & Options" paper for their (third) December 14–15, 1999 meeting in San Francisco, California. One of the proposals, submitted by Senator Ernest "Fritz" Hollings (D-SC), recommended a 5% *national sales tax* on Internet sales and other mail order and catalog sales that are not presently subject to sales taxes.[132]

A fourth (and last) meeting on March 20–21, 2000, in Dallas, Texas, preceded the presentation on April 1, 2000 of their recommendations to Congress concerning e-commerce taxation. The ACEC failed to reach the desired supermajority of 13 of the 19 members, but reached consensus on three issues:

(1) There should be no taxes on Internet access or usage,
(2) The 3% excise tax on telecommunications, dating back to the Spanish-American War, should be eliminated, and
(2) The existing system of sales taxation is far too complex.

Estimates of the number of state and local tax jurisdictions in the U.S. have ranged from 7,500 to 30,000, depending on the operational definition used. Historically, approximately 650 of these jurisdictions have changed or added new sales taxes in any given year. Therefore, the administrative burden is enormous. Despite the pre-existence and general availability of sales tax software by numerous vendors, many are concerned that sales tax compliance may inhibit growth through barriers to entry for start-up operations considering this new, electronic medium of commerce.

SUMMARY

Senator John McCain (R-AZ) campaigned for the Republican nomination for President through early 2000. He felt that the ITFA's three-year moratorium should be made permanent. As this monograph goes to press, a five-year extension of the ITFA has been approved by the U.S. House of Representative (H.R. 3709).

APPENDIX 5L
"TAX BRACKET CREEP" AND
INFLATION-INDEXING (1980s–)

Inflation is sometimes described as a tax on the money holders (Sennholz, 1976, 48).

That inflation makes the fair distribution of tax burdens more difficult is clear to all. How and whether to correct the distortions caused by inflation – that is, how and whether to index the tax system – is far less obvious.

Inflation distorts both the rate structure and the tax base. Discretionary changes in tax laws enacted by Congress have more than offset the tendency of inflation to push wage earners into higher rate brackets. They have had a weaker offsetting effect, however, on recipients of income from capital. The principal distortions affect the tax base; at the onset or during an acceleration of inflation, inflation-adjusted business income would probably be higher than income as currently calculated, but it would be lower in the long run (Aaron, 1976, vii).

INTRODUCTION

For 1974, the U.S. experienced double-digit inflation. Again, during the late 1970s and early 1980s the U.S. experienced three consecutive years of double-digit inflation, as measured by the U.S. consumer price index (CPI) (see 1974 and 1979, 1980, and 1981 in Appendix Table 5L-1 and Appendix Fig. 5L-1). As a result of this contemporary inflationary period, the issue of "tax bracket creep" received heightened interest.

The Canadian economy experienced comparable increases in inflation, though less severe. Since 1950, 1981 was the only year where Canada experienced double-digit inflation (see Appendix Table 5L-2 and Appendix Fig. 5L-2).

The Canadian and U.S. economies are related. The correlation coefficient for 1915–1998 for U.S./Canadian CPI-based measures contained in Appendix Tables 5L-1 and 5L-2 is 0.821 (p < 0.0001).

INFLATION-INDEXING IN CANADA AND THE U.S.

Since 1973, the personal income tax system has been indexed annually by raising the tax brackets and increasing the personal exemptions by an inflation factor based on CPI . . . (a)s an anti-inflation measures . . . the indexing rate was fixed at 6% for 1983 and 5% for 1984. Indexing . . . resumed in 1985. The indexing factor for the 1986 and subsequent taxation years is limited to the inflation rate less 3 percentage points. The personal exemption . . . ceased to be indexed in 1984 . . . indexed personal exemptions and deductions are rounded to the nearest $10 (Canadian Tax Foundation, 1987–1988, 7:6, referring to the Canadian income tax system).

. . . for 1986 and subsequent years, indexation is only for the amount of inflation in excess of 3%. Because of the continuing low level of inflation, there was no indexation for either 1993 or 1994 (Byrd, Chen & Jacobs, 1994–1995, 327, referring to the Canadian income tax system).

Beginning with the 1973 and subsequent tax years, Canada inflation-indexed their personal exemption and dependent allowances (see Appendix Table 6-2), as well as their tax brackets. The U.S. did not begin inflation-indexing the personal exemption and standard deduction until 1985 and subsequent years (see Appendix Table 6-1).

Presently, the U.S. inflation-indexes personal exemptions (see Appendix Table 6-1) and personal exemption phase-outs for AGI ranges (see Appendix Table 2-1), standard deductions (applicable only for non-itemizer taxpayers; see Appendix Table 6-1), wages bases for the Social Security component of FICA and SECA taxes (see Appendix Table 11-4), federal income tax brackets, and so on.

Historically, inflation has made it easier for governments to raise taxes, redistributing the tax burden from labor (e.g. wages, salaries, and self-employment earnings) to property income (in the form of *inflated* capital gains, upon realization) (Flemming, 1977, 90). This represents one of the arguments against the taxation of capital gains. Capital gains (and losses) are discussed further in Chapter 9.

SUMMARY

Inflation is a tax. To reduce or eliminate what was once referred to as "tax bracket creep", both Canada (1973–) and the U.S. (1985–) have employed inflation-indexing. However, Canada only resorts to inflation-indexation for measures in excess of 3% per year. The Canadian system, in effect, allows for a ceiling of 3% "tax bracket creep" per year. The U.S. has taken a more aggressive stance, attempting to completely eliminate the ordinary income component of "tax bracket creep".

APPENDIX 8M
FAMILY POLICY – HISTORICAL, CONTEMPORARY AND PROSPECTIVE EVIDENCE AND PERSPECTIVES

> ... ten rather extensive budget studies can be utilized to determine whether family unit costs change with changing family size. In all ten studies, budgetary evidence supports the contention that large families enjoy economics of scale and experience decreasing unit costs ... (Hansen, 1959, 126, as cited in Groves, 1963, 28–29).

INTRODUCTION

The economies of scale enjoyed by large families have received little or no direct attention or examination in U.S. individual tax policy discussions. However, U.S. family tax policy remains an important issue. For over two decades, family tax subsidies have been delivered through significant increases in the earned income tax credit (discussed in Chapter 7). We expect this trend to continue.

This chapter provides some international examples of the evolution of family tax policy. It includes some U.S. demographic information and trends, in the form of age distributions.

INTERNATIONAL EVIDENCE FROM THE 1970s

Kamerman and Kahn (1978) edited a collection of fourteen mid-1970s international papers on family policy issues (1).[133] A quarter of a century ago, the following nations were placed into three categories by Kamerman and Kahn:

- Those with an explicit, comprehensive family policy:

  (1) France – Formed official commissions in 1902 and 1912 to survey the depopulation problem, but policy did not significantly favor the family until the 1938–1946 (post-WWII) period.[134]

      Through the mid-1990s, the unit of taxation in France has remained the *family*. However, the French system of *'quotient familial'* (also applied in Luxembourg) takes into account the number of dependents. Taxable income of the entire family is divided by a coefficient, then the tax rate is applied (Messere, 1998, 13 and 100).
  (2) Sweden – Motivated by economic depression, unemployment, and the threat of a dwindling population and to promote a higher birth rate (1930s–).

      Traditionally, joint assessment of the income of husband and wife was used, but tax-motivated divorces to avoid the marriage tax penalty (associated with steeply progressive tax rates) led to taxation of the individual. Through the mid-1990s, the unit of taxation in Sweden, like Canada, remains the *individual* (Messere, 1998, 13 and 331).
  (3) Norway – Influenced by ideas and discussions originating in Sweden, but not until the post-WWII period.
  (4) Hungary – Introduced their first Family Act in 1952.
  (5) Czechoslovakia – A function of their post-World War II selection of socialism and gender equality (e.g. relatively high labor force participation for women, at 86.5% in 1975, compared to 96.8% for men).

- Those with an explicit, but narrowly focused family policy:

  (6) Austria – *Family policy* pre-dated WWII and provided *first marriage* cash payments, family allowance installments,[135] and tax reductions.
  (7) The Federal Republic of Germany (F.R.G.) – Following the totalitarian atmosphere of Nazi Germany, the family was the only social system left intact after WWII.

      Through the mid-1990s, the unit of taxation in Germany remains *optional*; married couples may choose to be taxed jointly or separately, with married couples benefiting from election of the joint filing option (Messere, 1998, 13 and 135).
  (8) Poland – Following Nazi occupation (1939–1945), the post-WWII changes in the socialistic political, social, and economic order rendered their old philosophy of helping the "poor" useless.[136]

(9) Finland – Concerns over declining birth rates and adverse long-term affects on the economy (1920s–) led to general alarm and a focus on decreasing death rates. They were greatly influenced by and followed the policies implemented in Sweden.

(10) Denmark – With a population of 3.8 million in 1940 and 5.0 million in 1975, the birth rate was declining in 1975; the majority of family policy measures are financed from national taxation.

- Those without an explicit family policy and where the notion is rejected:

(11) The United Kingdom (U.K.) – Family values, per se, have never been stressed in English public life, and the British government did not experience the concerns of declining birth rates or subsequent pronatalist policies implemented by Sweden, France, Belgium, Italy, and Germany.

    In recent years, fewer than 10% of the adult population filled in annual tax returns (Messere, 1998, 358).

(12) Canada – Taxing the *individual*, monthly family child-care allowances continue (1945–) as a post-WWII adjustment and economic stabilization measure and not as a deliberate policy to strengthen the family.[137] The objective for the French province of Quebec related to a positive population growth policy. The *individual* remains the unit of taxation in Canada.

(13) Israel – The founders of the kibbutz settlement were socialists from Eastern Europe and Russia, with a belief in greater equality between the sexes and viewing the family as an obstacle to the ideal of the community and a collective socialist society. Universal child allowances were first established in 1959.[138]

(14) The United States (U.S.) – Taxing the *family unit*, policies are not "harmonized" and contain many contradictions and discrepancies.[139] The unit of taxation in the U.S. remains the *family*, providing for the option of "joint" or "separate" filing for married taxpayers.

Kamerman and Kahn (1978, 483) identified *income-transfer* programs as the most ". . . readily identifiable instrument of family policy", with a system of family allowances as the most direct and efficient instrument, absent only in the U.S. *Dependency exemptions* in the individual income tax system represent a Western form of family policy, but favors taxpayers at the higher marginal Federal income tax brackets. They suggested that a change to uniform *tax credits* would favor poor large families (484). This has been the trend in the U.S. in recent decades.

INTERNATIONAL EVIDENCE FIFTEEN YEARS LATER (1989 TAX YEAR)

Pechman and Engelhardt (1990, 2), in an examination of the tax treatment of the family, conducted a comparative analysis of eleven countries (Australia, Canada, Denmark, France, Germany, Italy, Japan, Netherlands, Sweden, United Kingdom, and the United States) for the 1989 tax year. Some summary information is presented in Appendix Table 8M-1. They found the principle methods used to allocate tax burdens among family units to be:

(1) Personal exemptions, zero-bracket amounts, or tax credits;
(2) Head of household provisions;
(3) Deductions;
(4) Joint or separate filing; and
(5) Single or multiple tax rate schedules.

All of the above are present in the U.S. individual federal income tax system. First, personal exemption (see Chapter 6) deductions are available, as are tax credits for dependents. The contemporary U.S. standard deduction for non-itemizer taxpayers is comparable to its zero bracket amount predecessor. Second, the U.S. head of household is provided with increased standard deductions (for non-itemizers) and preferential tax rates schedules. Third, deductions (in the form of personal exemptions) are provided for dependents (as well as dependent deductions for medical and dental expenses). Fourth, the U.S. system provides married taxpayers with the option of joint or separate filing.[140] Fifth, the U.S. provides for separate tax rates schedules for single, married, married filing separately, surviving spouse, and head of household taxpayers.

CONTEMPORARY INTERNATIONAL TRENDS

Through the mid-1990s, the unit of taxation has been the *individual* for Canada, Italy, Japan, Netherlands, Sweden, and the United Kingdom. In recent decades, the United Kingdom has modified its position from one that permitted the *option* of joint or separate reporting to that of requiring *individual* reporting and taxation. This *option* has remained constant in Germany. Spain has moved from family taxation to providing the taxpayer(s) with the *option* of joint or separate reporting. The *family* has remained the unit of taxation in France. The United Stated continues to provide for taxation at the *family* unit level, through optional joint or separate filing for married taxpayers (Messere, 1998, 13).

Some of the summary information provided by Messere is provided in Appendix Table 8M-1.

The international trend appears to be toward establishment of the *individual* as the taxpaying unit, with France and the United States as the exceptions. We expect these exceptions to continue. The French system of *'quotient familial'* and the U.S. system of the EITC (discussed in Chapter 7) provide for comparable solutions to the support of the family from tax revenues.

SUMMARY

Historically, those nations providing the greatest economic incentives have been concerned with declining birth rates. Table 8M-1 summarizes some of the trends taking place in the U.S., with respect to age distribution.

Measures summarized in Table 8M-1 suggest a significant shift between the under-35 (a decline of 5.8%) and over-34 (an increase of 5.5%) age groups

Table 8M-1.   U.S. Population: Age Distribution (in millions)[1].

Year	< 17	18–24	25–34	35–44	45–54	55–64	65–74	> 74	Total
1960	64.5	16.2	22.9	24.2	20.6	15.6	11.1	5.6	180.7
Pct	*35.7%*	9.0%	12.7%	13.4%	11.4%	8.6%	6.1%	*3.1%*	
1970	69.8	24.7	25.3	23.2	23.3	18.7	12.5	7.6	205.1
Pct	34.0%	12.0%	12.3%	11.3%	11.4%	9.1%	6.1%	3.7%	
1980	63.7	30.3	37.6	25.8	22.8	21.8	15.7	10.1	227.8
Pct	28.0%	13.3%	16.5%	11.3%	10.0%	9.6%	6.9%	4.4%	
1990[2]	64.0	26.1	43.9	37.9	25.5	21.4	18.4	13.2	250.4
Pct	25.6%	10.4%	17.5%	15.1%	10.2%	8.5%	7.3%	5.3%	
2000[2]	65.7	25.2	37.1	43.9	37.2	24.2	18.2	16.6	268.3
Pct	*24.5%*	9.4%	13.8%	16.4%	13.9%	9.0%	6.8%	*6.2%*	
1990 to 2000 changes:									
Pct		*–5.8%*			*5.5% +*		*–0.5%*	*0.9% +*	
1960–2000 changes:									
Pct	*–11.2%*			*–8.1%*				*3.1% +*	

[1] *Source*: Statistical Abstract of the United States 1988 (108th Edition), U.S. Department of Commerce, Bureau of the Census (13 & 15).
[2] Estimated.

from 1990 to 2000, as the U.S. population ages. When comparing 1960 to 2000, this shift is far more dramatic. Therefore, in this respect, the U.S. would appear to be facing a declining birth rate and related concerns of continued financing of the Social Security system (discussed in Chapter 11). We would expect that the earned income tax credit would remain the preferred mechanism for providing economic incentives or relieving the financial hardships experienced by young families.

APPENDIX 10N

DAY TRADERS, BULL MARKETS, CAPITAL GAINS, AND THE STOCK MARKET CRASH OF 2000

Some had difficulty getting the spot cash to pay the fourth installment of the income tax and had to jettison part of the cargo to do it (Haugen & Lakonishok, 1988, 73, citing a 1920 article by G. E. Seldon).

. . . the strongest (stock market) effect (is) for the combined April estimated tax payment and final payment/return filing month (Cataldo & Savage, 2000, 181).

Mr. Kirk . . . sold stocks on Thursday to meet a $24,000 margin call and to raise cash to pay a $30,000 tax bill (*Wall Street Journal* April 17, 2000, A3)

INTRODUCTION

On Friday, April 14th, 2000 the U.S. stock markets "crashed". On April 17th, 2000 – the following Monday – a modest correction occurred. Figure 10N-1 illustrates the incredible rise in the NASDAQ (National Association of Securities Dealers Automated Quotation system), as well as the relatively stable Dow Jones Industrial Average (DJIA) and Standard and Poor's 500 Composite Index (S&P500) values, over twelve month period from July, 1999–June 2000.

Close-to-close declines on the NASDAQ, the DJIA, and the broad-based S&P 500 declined from 6.6% to 9.7% on this single trading day. The NASDAQ experienced a decline of 25.3% for the week ending Friday, April 14th, 2000. This decline was the result of: (1) high (0.7%) CPI-based inflation measures for the month of March, (2) related fears of Federal Reserve increases in interest rates to combat inflation, and (3) a self-perpetuating downward spiral from selling to meet margin calls for inexperienced day traders. This Friday also represented the last trading day for the week immediately preceding the April 17th, 2000 federal income tax return filing/payment for 1999 taxes and the

Fig. 10N-1. The NASDAQ "Crash" of April, 2000.

April 17th, 2000 first installment of quarterly estimated tax payments for the 2000 calendar/tax year, when *tax (estimated tax) payment effects* predict a decline (Cataldo & Savage, 2000, 181).

THE EARLY 2000 ECONOMIC ENVIRONMENT

In the months leading up to this stock market decline, Alan Greenspan, Chairman of the Federal Reserve, had expressed concerns about the "wealth effect". This effect is characterized by increased current consumption and the increase in demand-pull-based inflationary pressures caused by this increase in consumption. It results from individual investor efforts to convert anticipated long-term wealth increases, from increased values of retirement and deferred compensation-based stock portfolios, to current consumption.

The 1999 calendar year had resulted in the generation of large capital gains realizations. These gains resulted in higher 1999 calendar year taxes and larger first quarter 2000 calendar year estimated tax payments.

Figure 10N-2 illustrates how *tax (estimated tax) payment effects* may have contributed to the decline/correction sequence in mid-April, 2000. Historical evidence (Cataldo & Savage, 2000, 181) suggests that tax payments will occur on the business/trading day immediately preceding the day of payment (e.g. April 14, 2000). A correction occurs on the following business/trading day (e.g. April 17, 2000).

April 14, 2000 and April 17, 2000

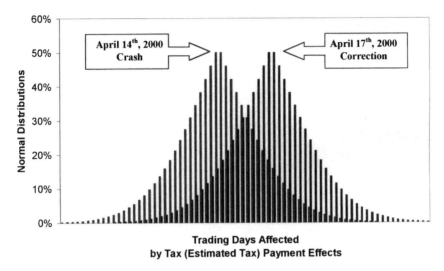

Fig. 10N-2. Normal Distribution Curve and the "Crash" of April 14, 2000.

Of course, all taxpayers will not wait until the last minute to liquidate marketable equity securities to pay their tax bill. Some will liquidate these assets and pay their taxes early. Others will miss the deadlines. However, as the normal distribution curve suggests and as all event-based studies presume, the majority will take action on these historically identified event dates.

SUMMARY

The rising stock market has: (1) resulted in the generation of taxable gains (1999) and increased U.S. revenues and budget surpluses, and (2) prompted policy-makers to propose that a portion of employer and employee contributions to Social Security taxes (the topic of Chapter 11) should be placed in the stock market, possibly under the control of the taxpayers themselves. However, we believe that the volatility of the NASDAQ and the uncertainties associated with a stock market, including day-trading, will require any proposals to either be too restrictive to have any long-term benefits and/or will simply not generate sufficient support to be legislated into law.

APPENDIX 11O
A U.S./CANADIAN COMPARISON OF TAXATION OF THE SELF-EMPLOYED

INTRODUCTION

The U.S. imposes a SECA tax on self-employed taxpayers. Canada does not. During the 1980s, the U.S. imposed even greater burdens on the self-employed taxpayer. How do the profits reported by U.S. and Canadian sole proprietorships/small businesses compare over this period?

A PRELIMINARY EXAMINATION OF THE U.S. AND CANADA

As the statistical evidence contained in Chapter 11 suggested, U.S. small business/sole proprietorship formation/retention and the percentage of sole proprietors reporting taxable profits declined over a period when U.S. SECA tax rates were increased. Table 11O-1 contains aggregate measures from both U.S. and Canadian tax returns data for the 1950–1996 tax years. An exploratory examination into these "net" measures of aggregate net income for the entire period (1950–1996), as well as the: (1) aggregate net income, (2) average net income, and (3) number of returns reporting self-employment or small business income for the 1979–1996 period, consistently produce Pearson correlation coefficients in excess of 90%.

SUMMARY

The U.S. has imposed a SECA tax on self-employed taxpayers since 1951. Canada has no SECA tax counterpart. Chapter 11 contains statistical evidence which suggests that U.S. taxpayers may have employed tax planning strategies to reduce self-employment net income, and the rising SECA taxes imposed, during a period when U.S. SECA tax rates rose and U.S. Federal income tax

Table 110-1. A U.S./Canadian (CAN) Comparison[1] of Aggregate Taxable
Net Income,[2] Average Taxable Net Income, and Number or Returns Filed for
Sole Proprietors[3] (1950–1996).

Tax Year	Aggregate NI		Average NI		No. Returns		Sources: *IRS SOI* & *DNR/RC TS*, respectively
	U.S.	CAN	U.S.	CAN	U.S.	CAN	
1950	$15B	$1M	NA	NA	NA	NA	1950, 39 & 1952, 83
1951	$16B	$2M	NA	NA	NA	NA	1951, 29 & 1953, 35
1952	$16B	$2M	NA	NA	NA	NA	1952, 21 & 1954, 35
1953	$17B	$2M	NA	NA	NA	NA	1953, 26 & 1955, 35
1954	$17B	$2M	NA	NA	NA	NA	1954, 36 & 1956, 35
1955	$18B	$2M	NA	NA	NA	NA	1955, 21 & 1957, 35
1956	$21B	$2M	NA	NA	NA	NA	1956, 21 & 1958, 36
1957	$20B	$2M	NA	NA	NA	NA	1957, 21 & 1959, 40
1958	$21B	$2M	NA	NA	NA	NA	1958, 28 & 1960, 39
1959	$21B	$2M	NA	NA	NA	NA	1959, 25 & 1961, 39
1960	$21B	$2M	NA	NA	NA	NA	1960, 33 & 1962, 39
1961	$23B	$2M	NA	NA	NA	NA	1961, 34 & 1963, 41
1962	$24B	$2M	NA	NA	NA	NA	1962, 35 & 1964, 31
1963	$24B	$3M	NA	NA	NA	NA	1963, 33 & 1965, 33
1964	$26B	$3M	NA	NA	NA	NA	1964, 11 & 1966, 23
1965	$28B	$3M	NA	NA	NA	NA	1965, 12 & 1967, 25
1966	$30B	$3M	NA	NA	NA	NA	1966, 8 & 1968, 35
1967	$31B	$3M	NA	NA	NA	NA	1967, 10 & 1969, 35
1968	$32B	$3M	NA	NA	NA	NA	1968, 9 & 1970, 35
1969	$34B	$3M	NA	NA	NA	NA	1969, 13 & 1971, 37
1970	$33B	$3M	NA	NA	NA	NA	1970, 14 & 1972, 39
1971	$34B	$4M	NA	NA	NA	NA	1971, 16 & 1973, 39
1972	$39B	$5M	NA	NA	NA	NA	1972, 15 & 1974, 39
1973	$45B	$6M	NA	NA	NA	NA	1973, 14 & 1975, 39
1974	$44B	$7M	NA	NA	NA	NA	1974, 24 & 1976, 39
1975	$43B	$8M	NA	NA	NA	NA	1975, 25 & 1977, 39
1976	$48B	$8M	NA	NA	NA	NA	1976, 31 & 1978, 39
1977	$50B	$8M	NA	NA	NA	NA	1977, 22 & 1979, 39
1978	$57B	$9M	NA	NA	NA	NA	1978, 20 & 1980, 39
1979	$59B	$10M	$5.3K	$6.3K	11.2M	1.5M	1979, 19 & 1981, 57
1980	$53B	$10M	$4.6K	$6.2K	11.5M	1.6M	1980, 46 & 1982, 57
1981	$45B	$11M	$3.7K	$6.7K	12.2M	1.6M	1981, 37 & 1983, 59
1982	$41B	$11M	$3.2K	$6.8K	12.8M	1.6M	1982, 46 & 1984, 61
1983	$51B	$12M	$3.8K	$7.7K	13.4M	1.6M	1983, 13 & 1985, 61
1984	$58B	$14M	$4.1K	$8.3K	13.9M	1.7M	1984, 19 & 1986, 97
1985	$67B	$15M	$4.6K	$8.8K	14.5M	1.7M	1985, 21 & 1987, 101
1986	$83B	$16M	$5.6K	$9.1K	14.9M	1.8M	1986, 25 & 1988, 107
1987	$104B	$18M	$6.8K	$9.9K	15.4M	1.8M	1988, 2 & 1989, 107
1988	$125B	$20M	$7.9K	$10.6K	15.9M	1.9M	1988, 2 & 1990, 107

Table 11O-1. Continued.

Tax Year	Aggregate NI		Average NI		No. Returns		Sources: *IRS SOI* & *DNR/RC TS*, respectively
	U.S.	CAN	U.S.	CAN	U.S.	CAN	
1989	$133B	$21M	$8.0K	$11.1K	16.5M	1.9M	1991, 2 & 1991, 113
1990	$141B	$22M	$8.3K	$11.0K	17.0M	2.0M	1991, 2 & 1992, 119
1991	$139B	$21M	$8.0K	$10.5K	17.3M	2.0M	1991, 2 & 1993, 71
1992	$151B	$22M	$8.6K	$10.6K	17.6M	2.1M	1996, 2 & 1994, 61
1993	$152B	$23M	$8.5K	$10.6K	17.9M	2.2M	1996, 2 & 1995, 74
1994	$159B	$25M	$8.7K	$10.9K	18.2M	2.3M	1996, 2 & 1996, 54
1995	$161B	$26M	$8.8K	$11.0K	18.4M	2.4M	1996, 2 & 1997, 72
1996	$170B	$29M	$9.0K	$11.7K	18.9M	2.5M	1996, 2 & 1998, 72

Pearson correlation coefficient (r) for 1950–1996 (N = 47):

r = 0.973   (U.S.$_{\text{Aggregate NI}}$ × Canada$_{\text{Aggregate NI}}$)

Pearson correlation coefficient (r) for 1979–1996 (N = 18):

r = 0.969   (U.S.$_{\text{Aggregate NI}}$ × Canada$_{\text{Aggregate NI}}$)
r = 0.910   (U.S.$_{\text{Average NI}}$ × Canada$_{\text{Average NI}}$)
r = 0.932   (U.S.$_{\text{No. Returns}}$ × Canada$_{\text{No. Returns}}$)

[1] Where K = thousands (000 omitted), M = millions (000,000 omitted), and B = billions (000,000,000 omitted). NA = Not available (Canadian measures were not available for average NI and number of returns filed comparisons, therefore, neither U.S. nor Canadian measures are provided). All measures presented have been rounded, but non-rounded measures were used for statistical tests.

[2] Canadian measures in Canadian dollars and U.S. measures in U.S. dollars.

[3] Canadian measures include partnerships; U.S. measures exclude partnerships.

rates declined (1982–1990). The increases in SECA tax rates were imposed to improve the solvency of the U.S. Social Security system.

This Appendix contains evidence to suggest that comparable U.S. and Canadian self-employment/small business measures of: (1) aggregate net income, (2) average net income, and (3) the number of tax returns reporting small business or self-employment profits and losses were highly correlated (consistently above 90%) over 1979–1996 tax years. This period includes the 1982–1990 period. During this period the U.S. significantly increased SECA tax rates. Also during this period, Canada did not have (or impose or increase) a SECA tax counterpart.

Of course, one may argue, the U.S. and Canadian economies are intertwined. None-the-less, the exploratory evidence contained in Chapter 11, developed to examine the last U.S. Social Security tax increases (1982–1990), would suggest that a similar measure may be taken to resolve future concerns of U.S. Social Security insolvency. After all, what worked once before is likely to be tried again.

APPENDIX III

**(Supplemental Figures and Tables to
Appendix Chapters 2A–11O)**

Appendix Table 3C-1. Summary Measures Relating to Charitable Contributions (1917–1996).

Tax Year	Aggregate Charitable[1] Contributions	Percent of NI[2] or AGI[3]	Deductible Percent of NI[2] or AGI[3]	Source: *IRS SOI*
1917[4]	$245,080K	2.18992%	15%	1938, 66–70
1920[4]	$387,290K	1.63168%	15%	1938, 66–70
1922	$425,218K	1.99294%	15%	1938, 66–70
1923	$534,797K	2.15840%	15%	1938, 66–70
1924	$533,168K	2.07812%	15%	1938, 66–70
1925	$441,590K	2.01689%	15%	1938, 66–70
1926	$484,205K	2.20509%	15%	1938, 66–70
1927	$507,705K	2.25196%	15%	1938, 66–70
1928	$541,351K	2.10432%	15%	1938, 66–70
1929	$540,109K	2.09134%	15%	1938, 66–70
1930	$434,401K	2.20978%	15%	1938, 66–70
1931	$353,929K	2.27726%	15%	1938, 66–70
1932	$316,660K	2.41047%	15%	1938, 66–70
1933	$281,915K	2.32029%	15%	1938, 66–70
1934	$279,816K	2.11826%	15%	1938, 66–70
1935	$310,153K	2.02831%	15%	1938, 66–70
1936	$389,591K	1.99516%	15%	1938, 66–70
1937	$444,929K	2.06491%	15%	1938, 66–70
1938	$413,979K	2.15036%	15%	1938, 66–70
1939	$498,902K	2.20221%	15%	1939, 29–30
1940	$739,992K	2.05562%	15%	1940, 21–22
1941	$1,002,187K	1.72093%	15%	1941, 24–25
1942	$1,450,143K	1.84988%	15%	1942, 26–27
1943	$1,836,006K	1.85484%	15%	1943, 77 & 85
1944	$1,257,948K	1.08010%	15%	1944, 24–25
1945	$1,450,011K	1.20825%	15%	1945, 22–23
1946	$1,638,982K	1.22236%	15%	1946, 23
1947	$1,973,580K	1.31803%	15%	1947, 23–24
1948	$1,880,731K	1.15018%	15%	1948, 23
1949	$2,031,794K	1.26533%	15%	1949, 22
1950[4]	$2,260,342K	1.26171%	15%	1950, 42–45
1952	$3,116,483K	1.44757%	20%	1952, 9
1953	$3,552,448K	1.55327%	20%	1961, 180
1954[4]	$3,891,173K	1.69756%	30%	1961, 180
1956[4]	$4,877,793K	1.82195%	30%	1961, 180
1958[4]	$5,693,836K	2.02517%	30%	1961, 180
1960[4]	$6,750,326K	2.13980%	30%	1961, 180
1962[4]	$7,516,088K	2.15545%	30%	1962, 6 & 35
1964[4]	$8,327,000K	2.09927%	30%	1965, 210
1966[4]	$9,122,000K	1.94726%	30%	1967, 191

ANTHONY J. CATALDO II AND ARLINE A. SAVAGE

Appendix Table 3C-1. Continued.

Tax Year	Aggregate Charitable[1] Contributions	Percent of NI[2] or AGI[3]	Deductible Percent of NI[2] or AGI[3]	Source: *IRS SOI*
1968[4]	$11,138,925K	2.00910%	30%	1968, 6 & 65
1970[4]	$12,892,732K	2.04098%	30%	1970, 6 & 120
1972	$13,208,000K	1.77056%	30%	1972, 295
1973[4]	$13,895,720K	1.67995%	30%	1973, 4 & 53
1975	$15,393,331K	1.62413%	30%	1975, 1 & 48
1976	$16,792,387K	1.59336%	30%	1976, 2 & 58
1977	$17,266,462K	1.49042%	30%	1977, 1 & 51
1978	$19,691,249K	1.51186%	30%	1978, 9 & 56
1979	$22,210,838K	1.51569%	30%	1979, 2 & 36
1980	$25,809,608K	1.59937%	30%	1980, 2 & 57
1981	$26,445,456K	1.49189%	30%	1981, 2 & 54
1982	$33,471,694K	1.80719%	30%	1982, 2 & 61
1983	$37,677,955K	1.93957%	30%	1983, 2 & 51
1984	$42,119,812K	1.96830%	30%	1984, 2 & 59
1985	$47,962,848K	2.07995%	30%	1985, 2 & 62
1986	$53,815,979K	2.16852%	30%	1986, 2 & 65
1987	$49,623,907K	1.78900%	30%	1987, 2 & 50
1988	$50,949,273K	1.65257%	30%	1988, 2 & 49
1989	$55,459,205K	1.70310%	30%	1989, 2 & 41
1990	$57,242,757K	1.68093%	30%	1990, 2 & 40
1991	$60,573,565K	1.74839%	30%	1991, 2 & 45
1992	$63,843,281K	1.75918%	30%	1992, 37 & 76
1993	$68,354,293K	1.83583%	30%	1993, 31 & 78
1994	$70,544,542K	1.80535%	30%	1994, 79 & 1996, 3
1995	$74,991,519K	1.79005%	30%	1995, 60 & 1996, 3
1996	$85,159,305K	1.87742%	30%	1996, 60 & 1996, 3

[1] Includes individual returns with *net income*, including returns of estates and trusts (1916 though 1938), and individual returns with no *net income* (1928–1938). Fiduciary returns excluded for 1939–1943.

[2] From the 1913–1943 tax years, charitable contributions were calculated as a percentage of *net income* (see Fig. 3-1).

[3] For the post-1943 period, charitable contributions were calculated as a percentage of AGI (see Fig. 3-1).

[4] The IRS SOI does not provide detailed information for the 1918–1919, 1921, 1951, 1955, 1957, 1959, 1961, 1965, 1967, 1969, 1971, and 1974 tax years. These tax years represent breaks in the sequence of available information on charitable contributions.

Appendix Table 3C-2. Charitable Contributions

Matrix of Mann-Whitney U Tests for all possible combinations (noted if significant at the 10% level).

Deductible Percentage	15%	20%	30%	50%
15%		n.s.	n.s.	0.02
20%			0.05	0.06
30%				0.07
Tax Years =	1917–1951	1952–1953	1954–1969	1970–1996
N =	31	2	8	25
Mean =	1.9%	1.5%	2.0%	1.8%

N = 66 (1917–1996) at mean charitable contributions of 1.8% of *net income* (1917–1943) or the AGI comparable (1944–).

Appendix Table 3I-1.   The U.S. Passive Activity Loss (PAL) – Schedule E[1]
Rental Net Income/(Loss).

Tax Year	PALs Allowed[2]	Tax Returns with Rents	Percent Affected	Rental Amounts	Pct AGI	Source: *IRS SOI*
1980	100%	7,463,817	7.95%	$0.20B	+0.0%	1980, 43
1981	100%	7,778,048	8.15%	$–2.77B	–0.2%	1981, 40
1982	100%	7,795,010	8.18%	$–8.48B	–0.5%	1982, 49
1983	100%	8,085,300	8.39%	$–11.19B	–0.6%	1983, 17
1984	100%	8,493,465	8.54%	$–15.97B	–0.7%	1984, 22
1985	100%	8,932,714	8.79%	$–19.82B	–0.9%	1985, 24
1986	100%	8,803,495	8.54%	$–20.20B	–0.8%	1986, 29

TRA86 Imposes PAL Rules and Limits the Offset of pre-October
23, 1986, "Non-passive Income" by "Passive" Losses (PALs):

1987	65%	8,859,035	8.28%	$–15.76B	–0.6%	1987, 30
1988	40%	8,880,382	8.09%	$–11.86B	–0.4%	1988, 29
1989	20%	9,011,669	8.04%	$–9.39B	–0.3%	1989, 29
1990	10%	9,097,009	8.00%	$–7.56B	–0.2%	1990, 28
1991	–0–	9,126,578	7.95%	$–5.81B	–0.2%	1991, 32
1992	–0–	8,947,393	7.88%	$–0.38B	–0.0%	1992, 42
1993	–0–	9,039,688	7.89%	$3.27B	+0.1%	1993, 42
1994	–0–	9,157,881	7.90%	$6.05B	+0.2%	1994, 43
1995	–0–	10,224,313	8.65%	$17.19B	+0.4%	1995, 41
1996	–0–	9,188,523	7.63%	$9.89B	+0.2%	1996, 40

[1] The Form 1040, Schedule E includes rents, royalties, partnerships, estates, trusts, etc.
[2] Phase-out percentage of passive activity losses (PALs). Net loss includes non-deductible losses.

Appendix Table 3I-2. Pearson Correlation (r) Matrix for Descriptive
Statistics Contained in Appendix Table 3I-1.

	PCTALLOW	RENTRTRN	PCTRENT	RENTAMT	PCTAGI
PCTALLOW	0.000	0.165	0.390	**−0.650**	−0.740
RENTRTRN		0.000	**0.970**	−0.087	−0.139
PCTRENT			0.000	−0.223	−0.288
RENTAMT				0.000	**0.972**
PCTAGI					0.000

R-Square (R^2), Adjusted R-Square (adj-R^2),
and the C_p-Statistic of Reduced Model and Model Fit
as Applied to Regression Equation [10C]

Dependent Variable = RENTAMT

Model	Independent Variables		R^2	adj-R^2	C_p-Statistic
	PCTAGI	PCTALLOW			
1	X		94.4%	94.1%	4.8
2	X	42.2%	38.8%	190.1	
3	*X*	*X*	*95.5%*	*94.9%*	*3.0*

Statistical Results of Regression for Model 3

Dependent Variable = RENTAMT

	T-Statistic	P-Value	
Intercept	00.77	0.454	
PCTAGI	13.75	0.000	
PCTALLOW	1.94	0.070	
Overall F-Statistic		0.000	169.51
Adjusted R-Square			94.9%

294 ANTHONY J. CATALDO II AND ARLINE A. SAVAGE

Appendix Table 5L-1. U.S. Consumer Price Index-Urban (CPI-U)-Based Inflation Measures (1914–1998).

Year	CPI-U-Based Inflation	Year	CPI-U-Based Inflation	Year	CPI-U-Based Inflation	Year	CPI-U-Based Inflation
1914	1.0%	1936	1.5%	1958	2.8%	*1980*	*13.5%*[1]
1915	1.0%	1937	3.6%	1959	0.7%	*1981*	*10.3%*[1]
1916	7.9%	1938	−2.1%	1960	1.7%	1982	6.2%[1]
1917	17.4%	1939	−1.4%	1961	1.0%	1983	3.2%[1]
1918	18.0%	1940	0.7%	1962	1.0%	1984	4.3%[1]
1919	14.6%	1941	5.0%	1963	1.3%	1985	3.6%[1]
1920	15.6%	1942	10.9%	1964	1.3%	1986	1.9%[1]
1921	−10.5%	1943	6.1%	1965	1.6%	1987	3.6%[1]
1922	−6.1%	1944	1.7%	1966	2.9%	1988	4.1%[1]
1923	1.8%	1945	2.3%	1967	3.1%	1989	4.8%[1]
1924	−0-%	1946	8.3%	1968	4.2%	1990	5.4%[1]
1925	3.5%	1947	14.4%	1969	5.5%	1991	4.2%[1]
1926	1.1%	1948	8.1%	1970	5.7%	1992	3.0%[1]
1927	−1.7%	1949	−1.2%	1971	4.4%	1993	3.0%[1]
1928	−0-%	1950	1.3%	1972	3.2%	1994	2.6%[1]
1929	−1.7%	1951	7.9%	1973	6.2%	1995	2.8%[1]
1930	−2.3%	1952	1.9%	*1974*	*11.0%*	1996	3.0%[1]
1931	−9.0%	1953	0.8%	1975	9.1%[1]	1997	2.3%[1]
1932	−9.9%	1954	0.7%	1976	5.8%[1]	1998	1.6%[1]
1933	−5.1%	1955	−0.4%	1977	6.5%[1]		
1934	3.1%	1956	1.5%	1978	7.6%[1]		
1935	2.2%	1957	3.3%	*1979*	*11.3%*[1]		

[1] Beginning in 1975, Social Security beneficiaries have received annual cost-of-living allowance (COLA)/increases, since the system was linked to the CPI, as follows:

Always in *July*, for 1975 at 8.0%, 1976 at 6.4%, 1977 at 5.9%, 1978 at 6.5%, 1979 at 9.9%, 1980 at 14.3%, 1981 at 11.2%, 1982 at 7.4%, (there was no COLA for *six months* during 1983 to assist in bailing Social Security out of a financial crisis), always in *January*, for 1984 at 3.5%, 1985 at 3.5%, 1986 at 3.1%, 1987 at 1.3%, 1988 at 4.2%, 1989 at 4.0%, 1990 at 4.7%, 1991 at 5.4%, 1992 at 3.7%, 1993 at 3.0%, 1994 at 2.6%, 1995 at 2.8%, 1996 at 2.6%, 1997 at 2.9%, 1998 at 2.1%, 1999 at 1.3%, and 2000 at 2.4% (Associated Press 1999, A-15).

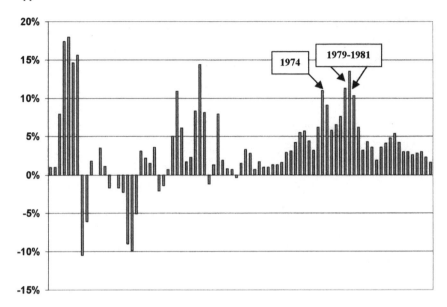

Fig. 5L-1. U.S. Consumer Price Index – Urban (CPI-U) (1914–1998).

Appendix Table 5L-2. Canadian[1] Consumer Price Index (CPI) – Based Inflation Measures (1915–1998).

Year	CPI-Based Inflation	Year	CPI-Based Inflation	Year	CPI-Based Inflation	Year	CPI-Based Inflation
1914	Unknown	1936	1.1%	1958	2.2%	1980	9.2%
1915	1.4%	1937	3.3%	1959	1.6%	*1981*	*11.0%*
1916	7.6%	1938	1.1%	1960	1.1%	1982	9.8%
1917	16.0%	1939	–0-%	1961	1.1%	1983	5.5%
1918	11.3%	1940	3.2%	1962	1.1%	1984	4.2%
1919	8.6%	1941	5.9%	1963	1.6%	1985	3.9%
1920	14.1%	1942	3.8%	1964	2.0%	1986	4.0%
1921	–13.4%	1943	1.9%	1965	2.0%	1987	4.2%
1922	–9.2%	1944	0.9%	1966	3.8%	1988	3.9%
1923	–0-%	1945	0.9%	1967	3.3%	1989	4.7%
1924	–1.9%	1946	2.7%	1968	4.0%	1990	4.6%
1925	1.8%	1947	8.9%	1969	4.3%	1991	5.3%
1926	–0-%	1948	12.1%	1970	3.3%	1992	1.5%
1927	–0.9%	1949	3.4%	1971	2.8%	1993	1.8%
1928	–0-%	1950	2.7%	1972	4.6%	1994	0.2%
1929	1.8%	1951	9.1%	1973	7.1%	1995	2.1%
1930	–0.9%	1952	3.0%	1974	9.6%	1996	1.6%
1931	–11.2%	1953	–1.2%	1975	9.9%	1997	1.6%
1932	–10.1%	1954	0.6%	1976	7.0%	1998	0.9%
1933	–4.7%	1955	–0-%	1977	7.3%		
1934	1.2%	1956	1.8%	1978	8.3%		
1935	1.1%	1957	2.8%	1979	8.4%		

[1] Source: *Statistics Canada Online Service*, 9J-RHC, Tunney's Pasture, Ottawa, Ontario K1A 0T6, Canada (http://www.statcan.ca January 1, 2000).

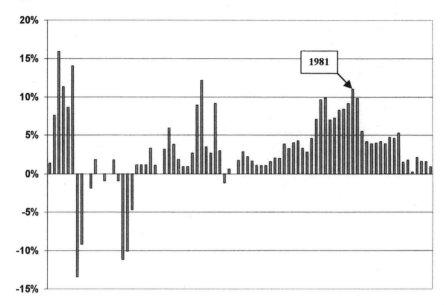

Fig. 5L-2. Canadian Consumer Price Index (CPI) (1915–1998).

Appendix Table 8M-1. Methods of Tax Burden Allocation (1989).

Country	(1) Personal Exemptions	(2) Head of Household	(3) Size/Rank of Deductions	(4) Unit of Taxation	(5) Rates Schedules
Australia	Zero-Bracket	Yes	None	Individual	Single
Canada	Tax Credit	Yes	None	Individual	Single
Denmark	Tax Credit	No	Seventh	Individual	Single
France	Zero-Bracket	No	Fourth	Household	Single
Germany	Zero-Bracket	Yes	Eighth	Household	Single
Italy	Tax Credit	Yes	Second	Individual	Single
Japan	Exemption	No	First	Individual	Single
Netherlands	Exemption	Yes	Fifth	Individual	Single
Sweden	Zero-Bracket	Yes	Sixth	Individual	Single
U.K.	Exemption	Yes	None	Individual	Single
U.S.	Exemption	Yes	Third	Household	Multiple

ANNOTATED BIBLIOGRAPHY

1. Boadway, R. W. and H. M. Kitchen. 1984. *Canadian Tax Policy (2nd Edition)*. Canadian Tax Paper No. 76. Canadian Tax Foundation; Toronto, Ontario.

   Contribution: Overview of the entire history of the Canadian tax system.

   As we have summarized, in Chapter 14, many recent changes in U.S. tax policy appear to represent the U.S. adoption of modifications or hybrids of Canadian tax policy mechanisms. This relatively early edition of Canadian tax policy issues chronicles the evolution of the entire Canadian tax system. The second section focuses on individual tax issues and is complemented by detailed tables of the entire history of the Canadian personal and dependency exemptions in its appendix (see Chapter 6 of this monograph).

2. Groves, H. M. 1963. Federal Tax Treatment of the Family. Studies in Government Finance. The Brookings Institution; Washington, D.C.

   Contribution: Still relevant discussion of issues and alternatives relating to the U.S. personal exemption deduction and family taxation.

   Groves provides a superior foundational discussion of family taxation issues, particularly those relating to the personal exemption deduction and alternatives, in the second section of his book. We gained many insights from his work (in the development of Chapter 6 of this monograph) and strongly recommend that those interested family taxation issues begin with this 1963 publication before reading more contemporary publications.

3. Joulfalan, D. 2000. *A Quarter Century of Estate Tax Reforms*. National Tax Journal (September).

   Contribution: In-depth and contemporary examination of the recent (past 25 years) historical evolution of the U.S. estate and gift tax.

   Chapter 13 provided discussion of the U.S. estate and gift tax. An excellent recent historical analysis of the last quarter century of estate and gift tax evolution is provided in this article. Issues addressed include: (1) estate and gift tax harmonization; (2) reductions in estates and gift taxes; (3) implemented administrative solutions to solve the liquidity problems associated with estate and gift taxes; and (4) ongoing compliance and administrative simplification efforts.

The author suggests that these estate and gift tax reforms have: (1) eliminated many taxpayers from estate and gift tax rolls; (2) deferred or reduced the applicability and burden of estate and gift taxes on married taxpayers; (3) reduced estate and gift taxes for businesses; and (4) reduced the yield of estate and gift taxes. Unintended effects include: (1) increased complexity, and (2) increased demand and the greater need for estate and gift tax planning.

4. Messere, K., ed. 1998. *The Tax System in Industrialized Countries*. Oxford University Press.

Contribution: In-depth and contemporary examination of the international tax systems, economic and social aspects, and recent tax reforms (and trends) for 10 industrialized countries.

Chapters 10 and 13 and Appendix 8M provided limited discussion of historical and contemporary international tax policies as they relate to capital gains, estate and gift taxation, and family tax policy, respectively. An excellent contemporary primer, the contents of this book were derived primarily from articles written in the early to mid-1990s and published in the mid-1990s in *Tax Notes International*.

This book examines the evolution of tax policy in 10 industrialized countries: (1) Canada, (2) France, (3) Germany, (4) Italy, (5) Japan, (6) the Netherlands, (7) Spain, (8) Sweden, the (9) United Kingdom (U.K.), and (10) the United States (U.S.). It follows the standardized framework established by Joseph Pechman in his book *Comparative Tax Systems: Europe, Canada and Japan* (Tax Analysts, 1987).

5. Perry, J. H. 1955. *Taxes, Tariffs, & Subsidies: A History of Canadian Fiscal Development*. Canadian Tax Foundation; University of Toronto Press.

Contribution: Perhaps the most detailed historical analysis of Canadian taxation, spanning three centuries.

Perry provides discussions of the history of Canadian taxation. He focuses, primarily, with the mid-1860s and chronicles developments through the mid-1950s. Appendix A of his book provides a three century summary of *Main Events in Taxation, 1650–1954*.

6. Pollack, S. D. 1996. *The Failure of U.S. Tax Policy: Revenue and Politics.* The Pennsylvania State University Press; University Park, Pennsylvania.

   Contribution: Political insights into historical U.S. tax policy formation.

   Pollack examines the history and the politics of legislation and judicial decisions surrounding the evolution of U.S. tax policy. Providing a convincing argument in favor of tax simplification, Pollack suggests that the flaw in the U.S. tax system stems from the fact that Americans charge their elected officials with conflicting *revenue-raising* and *political uses*. He accurately characterizes the contemporary U.S. tax system as a patchwork quilt. This characterization contributes to the lack of coherence in contemporary U.S. tax law. Pollack appears to be a proponent of simplified "flat" tax proposals.

   Pollack's work is insightful and enlightening, frequently providing a "behind the scenes" view of the political processes leading to tax policy formulation. This book would prove useful to those interested in further examination of Chapter 3 and related appendices. The citations are meticulous and provide the reader with a roadmap for further inquiry. Pollack's work represents an excellent primer as well as a source of reference for those initiating their study of the history of the U.S. system of taxation.

7. Shoven, J. B. and J. Whalley, Eds. 1992. *Canada-U.S. Tax Comparisons.* A National Bureau of Economic Research Project Paper; University of Chicago Press.

   Contribution: Discussions of the Pressures and Progress toward Harmonization of Income Tax Policy between Canada and the United States.

   Shoven and Whalley assembled and edited a wonderful selection of papers on the contemporary and ongoing efforts to coordinate free trade and tax policies between Canada and the U.S. Topics include Canadian and American reforms to Social Security systems, the Canadian national sales tax system, and numerous summaries and comparisons of the differences in tax policy. This publication will provide further background and supplement many of the chapters contained in this monograph (see the Chapter 14 summary chapter in this monograph for summary comparisons for Canada-U.S.).

ACRONYMS

ACEC	Advisory Commission on Electronic Commerce
ACRS	Accelerated Cost Recovery System
AFDC	Aid to Families with Dependent Children
AFT	Americans for Fair Taxation
AGI	Adjusted Gross Income
AICPA	American Institute of Certified Public Accountants
AMT	Alternative Minimum Tax (also see MT)
AMTI	Alternative Minimum Taxable Income
API	Atkinson-Plotnick index
AT	Alternative Tax
AZ	Arizona
BTA	Board of Tax Appeals
CA	California
CAN	Canada/Canadian
CBO	Congressional Budget Office
CCH	Commerce Clearing House
CG	Capital Gain (also see CL)
CL	Capital Loss (also see CG)
COBRA85	Consolidated Omnibus Budget Reconciliation Act of 1985
CPI	Consumer Price Index
CPI-U	Consumer Price Index-Urban
CPP	Canada Pension Plan (also see QPP)
CTF	Canadian Tax Foundation
CV	Coefficient of Variation
CYB	Canada Year Book
D	Democrat (also see R)
DE	Decomposition
DEFRA84	Deficit Reduction Act of 1984
DINK	Double-income, no kids
DJIA	Dow Jones Industrial Average
DNR	Department of National Revenue (Canada)

DTC	Department of Trade and Commerce (Canada)
DV	Dependent Variable (also see IV)
EI	Earned Income (also see EIC and EITC)
EIC	Earned Income Credit (also see EI and EITC)
EITC	Earned Income Tax Credit (also see EI and EIC)
ERISA74	Employee Retirement Income Security Act of 1974
ERTA81	Economic Recovery Tax Act of 1981
ESOP	Employee Stock Option Plan
ETR	Effective tax rate
FICA	Federal Insurance Contributions Act (also see SECA)
FIT	Federal Income Tax (also see SIT)
GATT	General Agreement on Tariffs and Trade
GNP	Gross National Product
GST	Goods and Services Tax (also see HST, MST, NRST, PST, and RST)
HE	Horizontal Equity (also see HI)
HH	Head of Household (filing status; also see MFJ, MFS, and SS)
HI	Horizontal Inequity (also see HE)
HR	House of Representatives (also see S)
HST	Harmonized Sales Tax (also see GST, MST, and PST)
IRA	Individual Retirement Account (U.S.; also see RRSP for the Canadian counterpart)
IRC	Internal Revenue Code (U.S.; see ITA for the Canadian counterpart)
IRS	Internal Revenue Service (U.S.; see RC for the Canadian counterpart)
ITA	Income Tax Act (Canada; also see IRC for the U.S. counterpart)
ITC	Investment Tax Credit
ITFA	Internet Tax Freedom Act
IV	Independent Variable (also see DV)
JCT	Joint Committee on Taxation
K	Thousands, symbol for
LT	Long-term (also see ST)
LTCG	Long-term Capital Gain
LTCL	Long-term Capital Loss
M	Millions, symbol for
MAXTAX	Maximum Tax

MFJ	Married, Filing Jointly (filing status; also see HH, MFS, and SS)
MFS	Married, Filing Separately (filing status; also see HH, MFJ, and SS)
MINTAX	Minimum Tax
MSA	Medical Savings Account (U.S.)
MST	Manufacturers' Sales Tax (also see GST, HST, and PST)
MT	Minimum Tax (also see AMT)
MTB	Marriage Tax Bonus (also see MTP)
M-T-D	Marriage-to-Divorce ratio
MTP	Marriage Tax Penalty (also see MTB)
NASDAQ	National Association of Securities Dealers Automated Quotation system
NI	Net Income
NOW	National Organization for Women
NRST	National Retail Sales tax (also see the Canadian GST)
OASDI	Old Age, Survivors, and Disability Insurance
OBRA87	Omnibus Budget Reconciliation Act of 1987
OBRA90	Omnibus Budget Reconciliation Act of 1990 (also see RRA90)
OBRA93	Omnibus Budget Reconciliation Act of 1992 (also see RRA93)
PAL	Passive Activity Loss
PL	Public Law
PST	Provincial Sales Tax (also see GST, HST and MST)
QPP	Quebec Pension Plan (also see CPP)
R	Republican (also see D)
RA78	Revenue Act of 1978
RC	Revenue Canada (Canadian; see IRS for the U.S. counterpart)
RDI	Rank Divergence index
RESP	Registered Education Savings Plan (Canada)
RIA	Research Institute of America
RP	Rank Preservation
RPP	Registered Pension Plan (Canada)
RRA89	Revenue Reconciliation Act of 1989
RRA90	Revenue Reconciliation Act of 1990 (also see OBRA90)
RRA93	Revenue Reconciliation Act of 1993 (also see OBRA93)

RRSP	Registered Retirement Savings Plan (Canada; also see IRA for the U.S. counterpart)
RST	Retail Sales Tax (also see Canadian GST)
S&P	Standard and Poor's 500 stock composite index
S	Senate (also see HR)
SC	South Carolina
SE	Self-Employed (also see SECA)
SECA	Self-Employment Contributions Act (also see FICA)
SEP	Simplified Employee Pension
SGL	Single
SIMPLE	Savings Incentive Match Plans for Employees
SIT	State Income Tax (also see FIT)
SOI	Statistics of Income (U.S.; see TS for the Canadian counterpart)
SSA	Social Security Administration
SS	Surviving Spouse (filing status; also see HH, MFJ, and MFS)
ST	Short-term (also see LT)
STCG	Short-term Capital Gain
STCL	Short-term Capital Loss
T	Trillions, symbol for
TAMRA88	The Technical and Miscellaneous Revenue Act of 1988
TEFRA82	Tax Equity and Fiscal Responsibility Act of 1982 (as amended by Technical Corrections Act of 1982)
TI	Taxable Income
TRA69	Tax Reform Act of 1969
TRA76	Tax Reform Act of 1976
TRA78	Tax Reform Act of 1978
TRA84	Tax Reform Act of 1984
TRA86	Tax Reform Act of 1986
TRA97	Taxpayer Relief Act of 1997 (as amended by the Tax Technical Corrections Act of 1997)
TS	Treasury Statistics (Canadian; see SOI for the United States counterpart)
UI	Unemployment Insurance (a Canadian component of individual taxation)
U.K.	United Kingdom
U.S.	United States
USA	Unlimited Savings Allowance
VAT	Value Added Tax

VE	Vertical Equity (also see VI)
VI	Vertical Inequity (also see VE)
WWI	World War I (also see WWII)
WWII	World War II (also see WWI)

GLOSSARY OF TERMS

ABILITY TO PAY is a tax policy concept, which suggests that taxpayers with higher (lower) incomes should pay a higher (lower) portion of their income economic resources in tax. The U.S. system of PROGRESSIVE TAXATION embraces this concept.

ABOVE-THE-LINE DEDUCTIONS (also see ADJUSTMENTS TO GROSS INCOME). This topic is covered in Appendix 2A.

ADJUSTED GROSS INCOME (AGI) represents the net result of GROSS INCOME less ABOVE-THE-LINE DEDUCTIONS or ADJUSTMENTS TO GROSS INCOME. It is the taxpayer's income before ITEMIZED DEDUCTIONS and PERSONAL EXEMPTIONS. This topic is covered in Appendix 2A.

ADJUSTMENTS TO GROSS INCOME are those deductions permitted taxpayers in arriving as ADJUSTED GROSS INCOME (AGI) (e.g. IRA contributions, alimony paid, one-half of self-employment taxes, etc.). This topic is covered in Appendix 2A.

ADVISORY COMMISSION ON ELECTRONIC COMMERCE (ACEC) was established in late 1999 to reach a consensus on the solution to Internet taxation. They were not successful in this effort, but made their results available to Congress in early 2000. The ACEC is discussed in Appendix 4K.

ALTERNATIVE MINIMUM TAX (AMT) is an extremely complex set of tax rules designed to prevent middle-income and wealthy taxpayers from completely escaping taxation through clever tax planning strategies.

AMERICAN INSTITUTE OF CERTIFIED PUBLIC ACCOUNTANTS (AICPA) is a national professional association for Certified Public Accountants (CPAs).

BUSINESS INTEREST is that interest paid or accrued on debts properly allocated to a trade or business. BUSINESS INTEREST is usually deductible.

CANADIAN PENSION PLAN (CPP) is the Canadian counterpart to the U.S. SOCIAL SECURITY system.

CAPITAL ASSET is property as defined (and not excluded) in Internal Revenue Code (IRC) Section 1221. CAPITAL ASSETS exclude stock in trade or inventory, realty and depreciable property used in a trade or business, certain intellectual property, accounts and notes payable evolving from trade or business endeavors, short-term government obligations issued at discount (pre-June 24, 1981), U.S. government publications that were not purchased, and property contemplated by the *Corn Products doctrine* or substituting for ordinary income.

CAPITAL GAIN is a gain from the actual or constructive sale or exchange of a CAPITAL ASSET.

CAPITAL LOSS is a loss from the actual or constructive sale or exchange of a CAPITAL ASSET.

CONSUMER PRICE INDEX (CPI) is a measure of inflation based on a typical consumer's "market basket of goods".

CONSUMPTION TAX is a tax levied on the consumption of some product or service. Because those at lower income levels consume higher percentages of their income, consumption taxes are classified as regressive. The sales tax is a form of consumption tax.

C_p-STATISTIC is used as one measure for the selection of the best-fitting model (i.e. not underspecified or overspecified – suggesting the presence of a multicollinearity problem). Theory suggests that the preferred model is one where the C_p-STATISTIC $\approx$ the number of parameters/independent variables, including the intercept.

DONEE is the recipient of a gift from a DONOR (see ESTATE TAX or GIFT TAX).

DONOR is the person who makes a gift to a DONEE (see ESTATE TAX or GIFT TAX).

EARNED INCOME represents income from wages or self-employment as opposed to that generated from capital.

EARNED INCOME (TAX) CREDIT (EI(T)C) is a refundable credit (e.g. workfare) made available to eligible individuals.

ELASTIC or ELASTICITY is the term used by economists in lieu of "responsiveness".

ESTATE TAX is a tax levied on the estate before any transfers. Also see INHERITANCE TAX.

FEDERAL INCOME TAX (FIT) is the individual tax imposed on U.S. taxpayers. This topic is covered in Appendix 2A.

FEDERAL INSURANCE CONTRIBUTIONS ACT (a.k.a., FICA) is a tax imposed on both employees and employers to provide funds for the SOCIAL SECURITY benefits of employees.

FICA (see FEDERAL INSURANCE CONTRIBUTIONS ACT).

FLAT TAX, also known as a PROPORTIONAL TAX, is a tax, which takes the same proportion of each taxpayer's income. Sales or consumption taxes are examples of a FLAT TAX.

GENERAL SALES TAX (GST) refers to a national sales tax applied in the Canadian tax system. Also see HARMONIZED SALES TAX and PROVIN-CIAL SALSE TAX.

GIFT TAX laws prevent ESTATE TAX or INHERITANCE TAX avoidance otherwise possible through the transfer of assets before death.

GROSS INCOME, for U.S. federal income tax purposes, is income from "any source derived". This topic is covered in Appendix 2A.

HARMONIZED SALES TAX (HST) refers to a combined national and provincial sales tax applied in the Canadian taxation system. Also see GENERAL SALES TAX and PROVINCIAL SALES TAX).

HEAD TAX is a tax on the very existence of some classified group of individuals.

HORIZONTAL EQUITY is a tax policy concept, which suggests that taxpayers in similar economic circumstances should bear similar tax burdens.

For example, there should be no difference in the taxation of long-term capital gains and salary income.

INCOME-AVERAGING was a technique used for the 1964 through 1986 tax years to reduce the inequitable affects of steeply progressive TAX RATES or BRACKETS on individual taxpayers with high fluctuations in TAXABLE INCOME. This topic is covered in Appendix 3E.

INCOME STATISTICS is the Canadian counterpart to the INTERNAL REVENUE SERVICE'S *Statistics of Income* publications.

INCOME TAX is a tax levied on the income of individuals, corporations, or trusts.

INDIVIDUAL RETIREMENT ACCOUNT (IRA) is a type of individual retirement arrangement, using a trust or custodial account and frequently providing the taxpayer with considerable flexibility in the selection of investments.

INFLATION-INDEXING represents the U.S. Government's effort to avoid the adverse effects of "tax bracket creep".

INHERITANCE TAX is a tax levied on individuals receiving property from the estate. Also see ESTATE TAX.

INTERNAL REVENUE CODE (IRC) is the primary source of U.S. taxes.

INTERNAL REVENUE SERVICE (IRS) is the largest administrative agency of the Treasury Department and is charged with the collection of U.S. taxes.

INTERNET TAX FREEDOM ACT (ITFA) is a piece of legislation deferring a decision on the national (sales) taxation of business-to-consumer sales originating from the Internet. This topic is covered in Appendix 4K.

INTER VIVOS means "during life", whereas, testamentary means "by will".

INTESTATE (see TESTATE) means "without a will".

ITEMIZED DEDUCTIONS are those deductions from ADJUSTED GROSS INCOME in arriving at TOTAL INCOME. This topic is covered in Appendix 2A.

INVESTMENT INTEREST is interest allowable as a deduction if allocable to property held for investment, but not otherwise qualifying for a deduction.

LOCK IN EFFECT is a term used to describe the tax-based incentive for taxpayers with appreciated property, including stocks and bonds, to avoid sales of these assets and the accompanying incidence of CAPITAL GAINS taxes.

MARRIAGE TAX is the name associated with the so-called "penalty" associated with the filing of joint returns in the U.S. For example, non-itemizing married taxpayers do not receive a STANDARD DEDUCTION representing double that amount available to single taxpayers.

MAXIMUM TAX rates were available (1971 through 1981), as a ceiling, until made obsolete by the reduction of maximum marginal tax rates on EARNED INCOME. This topic is covered in Appendix 3F.

MEDICAL SAVINGS ACCOUNT (MSA) is a relatively recent measure intended to promote, through tax deductions and incentives, medical coverage by otherwise uncovered Americans. These plans are in their infancy, are exploratory, and have not been successful.

MEDICARE taxes fund health care for elderly, retired taxpayers. The current MEDICARE tax rate is 1.45%, each, for employer and employee and 2.9% (1.45% *multiplied by* 2) for the self-employed taxpayer. The MEDICARE tax is a FLAT TAX (from the perspective of the low- or moderate-income taxpayer) or a REGRESSIVE TAX (from the perspective of the high-income taxpayer).

NET ESTATE equals the gross estate less estate tax deductions.

NONQUALIFIED PLAN is a term used to refer to a deferred compensation plan that is not a "qualified" deferred compensation plan and offers only limited tax benefits (see QUALIFIED PLAN).

NORMAL TAX, at FLAT TAX rates, was applied during the early years of U.S. individual federal income taxation.

NORMATIVE ECONOMICS is an approach, based on value judgements, designed to formulate recommendations as to what *should* be done.

OLD-AGE, SURVIVORS, AND DISABILITY INSURANCE (OASDI) represents the primary portion of U.S. employment taxes. It is paid by employees and employers (FICA) or self-employed (SECA) taxpayers.

PASSIVE ACTIVITY INTEREST is the interest expense resulting from passive activities. Also see PASSIVE ACTIVITY LOSSES. This topic is covered in Appendices 3G and 3I.

PASSIVE ACTIVITY LOSSES (PALs) represent the aggregate, net losses from all passive activities, and are generally not permitted to be used to offset income generated from non-passive sources. This topic is covered in Appendix 3I.

PERSONAL EXEMPTIONS or DEDUCTIONS are available, subject to phase-out for high-income taxpayers, as a deduction in arriving at TAXABLE INCOME. This topic is covered in Appendix 2A.

PERSONAL INTEREST is that interest expense not associated with: (1) a business, (2) home mortgage, (3) passive activities, or (4) investments. PERSONAL INTEREST is not deductible. This topic is covered in Appendix 3G.

POSITIVE ECONOMICS is a scientific approach to analysis seeks to formulates hypotheses of the "If . . . then" variety, and is absent of value judgements or opinion.

PROGRESSIVE TAX is a tax whose rates increase as the amount subject to taxation increases. An illustration is provided in Fig. 2–1.

PROPERTY TAX is a tax levied on the ownership of some particular set of goods (e.g. state automobile and real estate valuation taxes).

PROPORTIONAL TAX is a FLAT TAX (e.g. sales taxes).

PROVINCE is the Canadian counterpart to a U.S. state. All Canadian provinces are listed and detailed in a map in Appendix Fig. 4–1.

PROVINCIAL SALES TAX (PST) refers to a Canadian sales tax applied in its provinces. Also see GENERAL SALES TAX and HARMONIZED SALES TAX.

QUALIFIED PLAN is a term used to refer to a qualified deferred compensation plan, which is a pension, profit-sharing, or stock bonus plan meeting the requirements of Internal Revenue Code Sections 401, 410, and 411. These plans provide for employer contribution deductibility (subject to ceilings) and the exclusion of contributions and related returns from the gross income of participant gross income (until actually received) in addition to limited preferential (e.g. five-year forward averaging) or capital gains treatment for lump sum distributions.

QUALIFIED RESIDENCE INTEREST is deductible as an itemized deduction, but is subject to a mortgage ceiling of $1.1 million. This topic is covered in Appendix 3G.

QUEBEC PENSION PLAN (QPP) is the Quebec (provincial) version of the CANADIAN PENSION PLAN (CPP).

REGRESSIVE TAX is a tax that more adversely affects taxpayers with lower incomes. For example, a sales tax is regressive, in that those at lower income levels must consume a higher percentage of their income. Alternatively, higher income taxpayers might avoid sales taxation on a higher percentage of their income by simply deferring consumption and increasing investment.

REGISTERED RETIREMENT SAVINGS PLAN (RRSP) is the Canadian counterpart to the U.S. INDIVIDUAL RETIREMENT ACCOUNT (IRA).

RESIDENCE INTEREST (see QUALIFIED RESIDENCE INTEREST).

RETAIL SALES TAX (RST) is the name associated with U.S. proposals and discussions of a national sales tax.

REVENUE CANADA is the Canadian counterpart to the U.S. INTERNAL REVENUE SERVICE.

SALES TAX is a form of CONSUMPTION TAX.

SELF-EMPLOYMENT CONTRIBUTIONS ACT (see SECA).

SECA (Self-Employment Contributions Act) is a term used to describe the self-employed taxpayer's counterpart to the FEDERAL INSURANCE CONTRIBUTIONS ACT.

SOCIAL SECURITY (see FICA for employees/employers or SECA for self-employed taxpayers).

STANDARD DEDUCTION is that amount of itemized deductions presumed by the IRS. If a taxpayer has ITEMIZED DEDUCTIONS in excess of their STANDARD DEDUCTION, they may chose to fill out and file a Form Schedule A with their individual federal income tax return. This topic is covered in Appendix 2A.

STATITISTICS OF INCOME (SOI) is a division of the INTERNAL REVENUE SERVICE charged with the responsibility of accumulating a presenting tax return data . . . . The Canadian counterpart is to the INCOME STATISTICS series of publications.

STEP UP in tax basis is achieved through the ESTATE TAX.

SURTAX, at PROGRESSIVE TAX rates, were applied to the income of individual taxpayers in the early years of individual Federal income taxation. This topic is covered in Chapter 3.

TARIFF is a tax imposed on the importation or exportation of certain goods and/or services.

TAX CREDITS represent those amounts available to reduce TAXABLE INCOME on a dollar-for-dollar basis. This topic is covered in Appendix 2A.

TAX EVASION is a term used to describe illegitimate or illegal tax planning strategies.

TAX RATE or TAX BRACKET refers to the marginal tax rate applying to the last dollar of TAXABLE INCOME. This topic is covered in Appendix 2A.

TAXABLE INCOME (TI) is GROSS INCOME less all statutory deductions. This topic is covered in Appendix 2A.

TESTATE or TESTATOR or TESTAMENTARY means by will (after death), as opposed to during life.

TOTAL INCOME (see GROSS INCOME). This topic is covered in Appendix 2A.

TRANSFER TAX is a tax levied on the TRANSFER of property from one owner to another.

UNIFIED TRANSFER TAX refers to the combined ESTATE TAX and GIFT TAX system applicable to gratuitous transfers of wealth by gift and at death.

UNIT OF TAXATION in the U.S. is the household, where joint or separate filing options are provided to U.S. taxpayers. In contract, the Canadian UNIT OF TAXATION is the individual.

UNLIMITED MARITAL DEDUCTION applies to ESTATE TAX returns. It provides that the marital deduction has no limit.

USER TAX is a tax levied on the user of some facility or service (e.g. road or bridge tolls).

VALUE ADDED TAX (VAT) is a tax levied at each point of exchange of goods or services from primary production to final consumption. It is levied on the difference between sales price of the outputs, to which the VAT is applied, and the cost of the inputs. The ultimate consumer bears the cost of the entire VAT. Also known as a cascade tax or turnover tax.

VERTICAL EQUITY is a tax policy concept, which suggests that the ABILITY TO PAY concept should apply to taxation (e.g. the greater the income, the higher the tax).

WAGE BASE is the term used to describe the ceiling to which FICA or SECA taxes are applied. An illustration is provided in Fig. 2-2.

NOTES

1. Old-age, survivors, disability, and hospital insurance, comprising both employer and employee contributions, approximated 32% of total U.S. federal income taxes collected for the 1996 tax year.

2. The Act of August 5, 1861, chap. 45, sec. 49, 12 Stat. 309 was never put into effect, but a later Act was signed into law by President Lincoln on July 1, 1862 (chap. 119, 12 Stat. 432, 469–471), providing for a maximum tax rate of 5%. It was amended by the Act of June 30, 1864, chap. 173, sec. 116, 13 Stat. 223, with a maximum tax rate of 10%.

3. From 1874–1984, approximately 68 bills were introduced to restore the income tax (Pollack, 1996, 47). In 1895, the Supreme Court held the 1894 tax to be an unconstitutional "direct" tax in the case of *Pollock v. Farmer's Loan and Trust Co.*, 157 U.S. 429c (1895); 158 U.S. 601 (1895) (rehearing).

4. Also known as the *Payne-Aldrich Tariff Act* of August 5, 1909, 36 Stat. 112.

5. Ratified on February 25, 1913.

6. Popularly known as the *Underwood Tariff Bill* (Blakely, 1916, 839).

7. Public Law (PL) 63–16, chap. 16, 38 Stat. 114, 166–181.

8. The *Revenue Act of 1914* (PL 63–217, 38 Stat. 745) raised excise taxes only.

9. PL 64–271, 39 Stat. 756 was enacted September 8, 1916, and amended March 3, 1917 by PL 64–377

10. PL 64–377, 39 Stat. 1000, was enacted on March 3, 1917. It was followed by American entry into World War I in April 1917, and PL 65–50, 40 Stat. 300 (the *War Tax Law*), enacted by Congress on October 3, 1917.

11. PL 65–254, 40 Stat. 1057, enacted February 24, 1919.

12. Many of these so-called War Taxes were later repealed by the *Revenue Act of 1921*.

13. Enacted November 23, 1921.

14. The EITC is discussed in Chapter 7.

15. Short-term and long-term holding periods have varied throughout history, and are discussed and summarized in Chapter 10.

16. This Act was amended by the Acts of March 13, 1924 and enacted June 2, 1924.

17. The EIC (or EITC) is discussed in Chapter 7.

18. Estate and gift taxation is discussed is Chapter 13.

19. Enacted February 26, 1926.

20. The contemporary Joint Committee on Taxation (JCT), initially known as the Joint Committee on Internal Revenue Taxation, was created by this Act.

21. PL 78–562, enacted May 29, 1928.

22. The Board of Tax Appeals (BTA) was established under this act. Initially an independent agency of the *executive* branch, it was renamed the Tax Court of the United States in 1942 and again renamed as the United States Tax Court in 1969.

23. Enacted June 6, 1932.

24. Enacted June 16, 1933.

25. PL 216, 48 Stat. 680, enacted May 10, 1934.

26. PL 407, 49 Stat. 1014, enacted August 30, 1935.

27. Enacted June 22, 1936.

28. PL 554, 52 Stat. 447, enacted May 28, 1938.

29. The first *alternative tax* is discussed in Chapter 9.

30. Also known as the *Internal Revenue Code of 1939* (IRC39), PL 155, 53 Stat. 862; enacted February 10, 1939.

31. The compensation of a judge taking office after June 6, 1932 was taxable beginning with the *Revenue Act of 1932*.

32. Enacted June 25, 1940.

33. Enacted September 20, 1941.

34. Enacted October 21, 1942.

35. PL 78–68, 57 Stat. 126, enacted June 9, 1943.

36. Enacted May 29, 1944.

37. PL 80–471, 62 Stat. 110, enacted April 2, 1948.

38. PL 82–183, 65 Stat. 452, enacted October 20, 1951.

39. The history of the self-employment tax is discussed and summarized in Chapter 11.

40. PL 83–591, 68A Stat. 3, also known as the *Internal Revenue Code of 1954* (IRC54).

41. PL 85–866, 72 Stat. 1606, Title 1.

42. PL 87–834, 76 Stat. 960.

43. PL 88–272, 78 Stat. 19.

44. PL 91–172, 83 Stat. 487, Titles 1–9.

45. Beginning with the 1982 tax year, the *Economic Recovery Tax Act of*

1981 (ERTA81) reduced maximum marginal individual federal income tax rates to the MAXTAX rate of 50%. ERTA81 made the MAXTAX obsolete.

46. PL 92–178, 85 Stat. 497.

47. PL 93–406, 88 Stat. 829.

48. PL 94–12, 89 Stat. 26.

49. PL 93–483, enacted in October 1974.

50. PL 94–455, 90 Stat. 1520.

51. The *Tax Reduction and Simplification Act of 1977* further extended the EIC through the 1978 tax year.

52. These rules are found at IRC section 465.

53. These rules applied to sales occurring after July 26, 1978. This amount was later increased to $125,000, beginning with the 1982 tax year.

54. PL 97–34, 95 Stat. 172.

55. PL 97–248, 96 Stat. 324.

56. PL 97–414, enacted January 4, 1983.

57. PL 97–424, enacted January 6, 1993.

58. PL 97–448, enacted January 12, 1983.

59. PL 97–473, enacted January 14, 1983.

60. PL 98–67, enacted August 5, 1983.

61. PL 98–76, enacted August 12, 1983.

62. PL 98–21, 97 Stat. 65, enacted April 20, 1983.

63. PL 98–369, 98 Stat. 494.

64. PL 99–514, 100 Stat. 2085.

65. PL 100–203, 101 Stat. 1330–282; also known as Titles 9 and 10 of the *Omnibus Budget Reconciliation Act of 1987* (OBRA87).

66. Part II of Title IX of the OBRA87.

67. PL 100–647, 102 Stat. 3342.

68. PL 101–239, 103 Stat. 2301, Title 7.

69. ESOPs were originally recognized as a qualified stock bonus plan under ERISA74. ESOPs had previously been expanded in 1976, 1984 and 1986.

70. PL 101–508, 104 Stat. 1388.

71. PL 103–66, 107 Stat. 312, enacted August 10, 1993.

72. PL 104–168, enacted July 30, 1996.

73. PL 104–191, enacted August 21, 1996.

74. PL 104–191, enacted August 21, 1996.

75. PL 104–193, enacted August 22, 1996.

76. PL 105–34, enacted August 5, 1997.

77. PL 105–206.

78. PL 105–178.

79. PL 105–277.

80. This refundable tax credit is the only component of the Canadian FIT system that is not tax neutral (e.g. results in a "marriage tax penalty/bonus"). The amount is modest.

81. Introduced by Pete Domenici and Sam Nunn in April 1995.

82. Designated as H. R. 2060 and S. 1050, and entitled the "Freedom and Fairness Restoration Act", this proposal was originally introduced in the 103rd Congress (June 16, 1994) and reintroduced in the 104th Congress (July 19, 1995). It provided for a flat 17% rate of tax, and was based on a 1981 plan by two Hoover Institution fellows (Iyer, Seetharaman & Englebrecht, 1996, 87–88).

83. Designated as S. 488, and entitled the "Flat Tax Act of 1995", this proposal is similar to the Armey-Shelby-Craig proposal, except that the statutory rate is 20% instead of 17% (Iyer, Seetharaman & Englebrecht, 1996, 88–89).

84. In an earlier work, Plotnick (1984) examined five "good" measures of horizontal inequity. He required three properties: (1) their values had to be independent of the final distribution of well-being; (2) they had to satisfy a simple anonymity condition; and (3) if one distribution differs from another in terms of rank-preservation, the index must show less horizontal inequity for the first redistribution. He used micro data for 1974 and found "no support for preferring one measure over the others" and that "all measures embody normative judgements" (29).

85. See Anderson (1985), where the alternative minimum tax (see Chapter 9), the capital gains tax (see Chapter 10), and passive activity losses (PALs) were examined (see Appendix 3I).

86. Appendix Table 2-1 summarized the more contemporary (1991–) U.S. phase-out ranges for inflation-indexed (1985–1986 and 1990–) personal exemption deductions.

87. Inflation-indexed "applicable amounts" were: $100,000, $50,000 if MFS (1991); $105,250, $52,625 if MFS (1992); $108,450, $54,225 if MFS (1993); $111,800, $55,900 if MFS (1994); $114,700, $57,350 if MFS (1995); $117,950, $58,975 if MFS (1996); $121,200, $60,600 if MFS (1997); and $124,500, $62,250 if MFS (1998).

88. Financial needs indexes were 80 for single taxpayers, 100 for married taxpayers, and 25 for each dependent (80:100:25). 1989 income tax exemptions plus the standard deduction was 55:100:200 (Pechman, 1987, 83).

89. MTPs include, but are not limited to, social security benefits (IRC Section 86), the earned income credit (IRC Section 32), the alternative minimum tax (IRC Section 55), individual retirement account contributions (IRC Section 219), net capital loss limitations (IRC Section 1211), the child

care credit (IRC Section 21), the credit for the elderly (IRC Section 22), moving expenses (IRC Section 217), once-in-a-lifetime personal residence gain exclusions for taxpayers over the age of 55 (IRC Section 121), the expense election (IRC Section 179), unemployment proceeds (IRC Section 85), restrictions regarding the election of differing accounting periods (IRC Section 6013), passive activity losses (IRC Section 469), the personal exemption (IRC Section 151) and/or itemized deduction (IRC Section 68) phase-outs. Property sales (IRC Section 267) and "wash" sales (IRC Section 1091) have also been identified by some as MTPs, but are associated with "related party" transactions and may involve aggressive tax planning fact patterns and assumptions. Similarly, medical deduction (IRC Section 213) limitations are AGI-based and may be perceived by some as a MTP. However, this topic is more appropriately suited to the tax planning options available to married taxpayers considering separate filing options (see Mullen, 1989).

90. Individual federal income tax returns for the 1939 tax year were the first to show the net short-term capital loss (STCL) carry-over provided by IRC Section 117(e) of the IRC. The net STCL realized for any tax year could not be deducted from income, but had to be carried forward and applied against future net short-term capital gain (STCG). This carry-over was restricted to one year (IRS SOI, 1939, 3–4).

91. The AMT was computed on Form 6251.

92. This alternative would: (1) raise tax revenues, (2) discourage lock-in for persons with short life expectancies, (3) stimulate savings, and (4) shift demand from growth stocks to income stocks.

93. For an analysis of several Social Security reform proposals, see Lyon and Stell (2000). They review three plans proposing the establishment of private accounts, as well as a fourth plan included in the Clinton Administration's FY 2001 budget. Their analyses, which ignored the possibility of an equity premium, suggests that all of these proposals resulted in *lower* net benefits for workers with average lifetime earnings.

94. The value of this deduction, as is the case with all deductions under U.S. Federal income tax law, is marginal FIT rate or tax bracket dependent. This is a result of the progressive tax rate system used in the U.S. For example, a one-hundred dollar deduction saves a U.S. taxpayer in the 15% FIT bracket fifteen dollars, but this same one-hundred dollar deduction would save a U.S. taxpayer in the 28% FIT bracket twenty-eight dollars. Most U.S. states also employ progressive tax rate structures.

95. Recall that this tax policy concept suggests that individuals in similar economic circumstances should bear similar taxes, regardless of the source or

form of the economic resources received. For example, there is no reason to impose different tax burdens on persons compensated with wages or self-employment earnings from their labor versus those generating gains from their use of capital in the stock market.

96. Generally, in the U.S., each of the states require separate income tax returns. All U.S. states do not impose an income tax. This procedure differs from that used in Canada, for example, where the provincial and Federal income tax is calculated and collected by Revenue Canada on a single, annual income tax return.

97. In pursuing the SECA tax reduction, the U.S. sole proprietor receives a tax benefit for every dollar of business expenses. The employee, though permitted to deduct what U.S. tax law refers to as "employee business expenses," must first exceed a specified minimum threshold before these business expenses result in any reduction in FIT liability. This threshold is specified as the excess of 2% of the taxpayer's AGI and, therefore, varies from individual taxpayer to individual taxpayer. For example, a U.S. taxpayer with an AGI of $10,000 and $500 of employee business expenses receives a tax deduction of only $200 or the amount in excess of 2% of AGI. Alternatively, all legal business expenses reduce taxable income for sole proprietors, partnerships, and corporations.

98. For the 1958–1980 tax years, married taxpayers were provided a maximum IRC Section 179 expense election of $4,000 per year. Single taxpayers were permitted to take a maximum expense election of half this amount ($2,000) over the same period. The measures used were the result of a marital status-dependent weighted average of these two expense election ceilings.

99. Depreciation expense is provided separately for the Schedule C only. A separate measure of depreciation is not provided for the Schedule F.

100. Interest expense is provided for the Schedule C only. A separate measure of interest expense in not provided for the Schedule F.

101. The exclusion of 4,495 records (see Table 11-2) compensates for the IRS practice of over representing high income taxpayers in their sampling procedure. Although it comprises 29.4% of the total returns contained in the sample, this component only represents 9.18% of the population (see Table 11-3).

102. Ceilings decreased from 70% in 1981, to 50% (1982–1986), 38.5% (1987), 33% (1988–1990), and 31% (1991 and 1992).

103. Married taxpayers, filing separate returns, were exposed to a phase-out at AGI levels ranging from zero to $10,000.

104. The U.S. estate tax is not an inheritance tax. The former is levied on an estate in its entirety; the latter is levied on the share received by the individual legatee or beneficiary (IRS SOI, 1921, 25).

105. The supplemental tax exemptions are not provided in this table, but are summarized in IRS SOI Bulletin (1996–1997, 11).

106. The term "unified" is used for both estate and gift tax purposes.

107. For tax years beginning after December 31, 1998, individual taxpayers eligible for Medicare could choose the traditional Medicare program or a Medicare + Choice MSA. The cut-off applying to MSAs does not apply to the Medicare + Choice MSA.

108. An eligible self-employed taxpayer claims deductions for contributions to MSAs (reported by the trustee on Form 5498-MSA). Individual taxpayers use Form 8853 to calculate MSA deductions and taxable income. Employer deductions to an MSA are excludable from the employee's income (and from wages for Social Security tax purposes). Distributions for medical expenses are tax-free (reported by the trustee on Form 1099-MSA).

109. A "small employer", for MSA purposes, is defined as an employer with 50 or fewer employees, but permitted to grow to as many as 200 employees.

110. A "high-deductible" plan was one with an annual deductible of $1,500 to $2,250 ($3,000 to $4,500) for individual taxpayers (families). The maximum annual contribution is 65 (75)% of the deductible for individual taxpayers (families). The maximum out-of-pocket expenses may not exceed $3,000 ($5,500) for individual taxpayers (families).

For the 1999 tax year, a "high-deductible" plan was one with an annual deductible of $1,550 to $2,300 ($3,050 to $4,600) for individual taxpayers (families). The maximum out-of-pocket expenses may be no more than $3,050 ($5,600) for individual taxpayers (families).

For the 2000 tax year, a "high-deductible" plan was one with an annual deductible of $1,550 to $2,350 ($3,100 to $4,650) for individual taxpayers (families). The maximum out-of-pocket expenses may be no more than $3,100 ($5,700) for individual taxpayers (families).

111. These deductions provide a tax benefit if the amounts for any given year exceed a specified threshold, presently established at 7.5% of AGI.

112. Under present U.S. tax law, the home mortgage interest deduction is limited to interest expenses associated with the first $1.1 million of principal residence financing.

113. Casualty and theft losses are reduced by $100 per incident and further reduced by 10% of the taxpayer's AGI.

114. Generally, miscellaneous itemized deductions must exceed a threshold

of 2% of AGI before the taxpayer achieves any tax savings/reduction as a result of this deduction.

115. The *basic* standard deduction amounts available for the entire history of the U.S. individual federal income tax, for both single taxpayers and married taxpayers filing joint returns, are discussed in Chapter 6 of this monograph.

116. These overpayments are typically made by relatively high-income employees with more that one employer within the tax year.

117. Unveiled on June 18, 1987, and modified and elaborated upon on December 16, 1987, draft legislation was made public on April 13, 1988; final legislation, Bill C-139, was read in the Canadian House of Commons on June 30, 1988 (Canadian Tax Foundation, 1988–1989, 7:1).

118. The 36% and 39.6% brackets, applicable in the U.S. only, are not reflected in Fig. 2B-1, due to the relatively low taxable income level of $100,000 (Canadian) which is less than $67,000 (U.S.).

119. For the 1917–1927 tax years, charitable contributions approximated 1.5% to 2.0% of net income (IRS SOI, 1927, 29–31).

120. For the 1945 tax year, charitable contributions approximated 4.1% of AGI.

121. PL 465–82, approved July 8, 1952.

122. For the 1954 (1958) tax year, 93 (96)% of the taxpayers itemizing deductions claimed the charitable contributions deduction (IRS SOI, 1954, 6 and 1958, 6, respectively).

123. Prior to the 1970 tax year individual taxpayers could, in certain cases, contribute and deduct unlimited amounts donated to charities. To qualify, contributions *plus* income tax had to exceed 90% of taxable income for eight of the 10 preceding tax years (IRC section 170). This unlimited charitable deduction was to be phased down to the 50% limitation over a five-year period (i.e. 80% for the 1970 tax year, reduced in 6% increments for the 1971–1974 tax years, until the 50% limitation is reached for the 1975 (and future) tax year(s) (IRS SOI, 1970, 280).

124. For the 1945 tax year, deductible medical and dental expenses for taxpayers able to itemize approximated 2.5% of AGI (IRS SOI, 1944, 27).

125. The maximum medical expense deduction was increased to $5,000 (from $2,500) for each exemption and the overall limitation was increased to $10,000 (from $5,000) for single taxpayers and $20,000 (from $10,000) for taxpayers filing joint returns, heads of household, and surviving spouses. For disabled taxpayers age 65 or older, the maximum deduction was increased to $20,000 each ($40,000 combined) (IRS SOI, 1962, 3).

126. This provision expired on July 1, 1992, but was extended by the Omnibus Budget Reconciliation Act of 1993, retroactively, through December

31, 1993. Therefore, taxpayers had to file amended returns to take advantage of the extension of this provision of tax law (IRS SOI, 1992, 114–115).

127. Contemporary tax law also provides for the deductibility of student loan interest, as an adjustment or above-the-line deduction. This contemporary deduction is not addressed further.

128. IRC section 469 limits deductions for passive activities.

129. AGI was not adjusted or reduced for this procedure.

130. PL 105–277, included as Titles XI and XII of the Omnibus Appropriations Act of 1998 and approved as H.R. 4328 by Congress on October 20, 1998.

131. H.R. 1054/S. 442 originally proposed a six-year moratorium on the taxation of e-commerce.

132. Entitled the Sales Tax Safety Net and Teacher Funding Act (S. 1433).

133. Some of this information has been updated, where noted, using Messere (1998).

134. The National Institute of Demographic Studies, in 1976, suggested that a couple's expenses rise 27% for one child, 54% for two children, 81% for three children, and 108% for four children.

The National Union of Family Associations produced results varying by age, using the 1975 cost of raising one child under six years of age as a base, and suggested that costs of a child at 7–10 years of age increases by 66%, at 11–14 years of age increases by 131%, and at 15–18 years of age increases by 306%.

Finally, social indicators relating to the Seventh National Plan (1976) produced an incremental scale, using one adult as the base, and suggested that the cost of maintenance for one additional adult at an additional 70%, one child under 2 years at an additional 20%, one child aged 2–9 at an additional 40%, and one child aged 10–15 at an additional 50%.

135. Using one child as a base, an additional 109% is provided for two children, an additional 240% is provided for three children, an additional 353% is provided for four children, and an additional 120% is provided for each additional child above four. Amounts for severely handicapped children amount to 200% of the base amount.

136. In 1975, low-income families were entitled to monthly allowances, based on the number of children in their household. Using one child as a base, the allowance increased by 156% for two children, the allowance increased by 369% for three children, and by 25% for each additional child.

137. Monthly benefit rates (1975–1976), in Canadian dollars, were $13.25 for the first child, $19.87 for the second child, $32.84 for the third child, and

$36.16 for the fourth and subsequent children. Quebec added an additional $3.68, $4.92, $6.14, and $7.36, per month, respectively.

138. In 1976, child allowances as a proportion of the average wage, was 4.4% for one child, 8.8% for two children, 17.6% for three children, 28% for four children, 38.3% for five children, 49.8% for six children, 61.2% for seven children, and 72.6% for eight children.

139. Hansen (1959) used four budget studies to develop an average relative cost pattern based on family size. Beginning with a single person and adding spouse and children, he arrived at an index of 70 (single), 100 (married without children), 128 (married with one child), 153 (married with two children), 174 (married with three children), and 195 (married with four children). "He suggests that a 2–4–1–1/2 ratio most closely fits the pattern of relative costs but would be willing to settle for a 2–4–1 pattern as representing a fair approximation (married couple twice single taxpayer and dependents half as much as single taxpayer) (Groves, 1963, 29).

140. Generally, it is not to the taxpayer's advantage of married taxpayers to file separately.

REFERENCES

Aaron, H. J. (Ed.) (1976). *Inflation and the Income Tax*, Studies of Government Finance. Washington, D.C.: The Brookings Institution.

Alm, J., & Whittington, L. A. (1993). Marriage and the Marriage Tax. *1992 Proceedings Of The Eighty-fifth Annual Conference On Taxation* (National Tax Association – Tax Institute Of America) pp. 200–205.

Alm, J., & Whittington, L. A. (1995). Does the Income Tax Affect Marital Decisions? *National Tax Journal, 48,*(4), 565–572.

Anderson, K. E. (1988). A Horizontal Equity Analysis of the Minimum Tax Provisions: 1976–1986 Tax Acts. *The Journal of the American Taxation Association, 10* (Fall), 6–25.

Anderson, K. E. (1985). A Horizontal Equity Analysis of the Minimum Tax Provisions: An Empirical Study. *The Accounting Review, 60* (July), 357–371.

Associated Press, The (1999). *Retirees to get Social Security raise*, as reprinted in The Oakland Press (Wednesday, October 20), A-15.

Auerbach, A. J., & Hassett, K. A. (1999). A New Measure of Horizontal Equity, *National Bureau of Economic Research*, No. 7035. Cambridge, MA.

Auster, R. (1988). Working with the $10,000 Expense Election. *TAXES-The Tax Magazine*, (July), 537–542.

Auten, G. E., & Reschovsky, A. (1997). The New Exclusion for Capital Gains on Principal Residences. *Proceedings of the 90th Annual Conference on Taxation* (November 9–11, 1997), 223–230.

Auten, G. E. (1983). Capital Gains: An Evaluation of the 1978 and 1981 Tax Cuts. In: C. E. Walker & M. A. Bloomfield (Eds), *New Directions in Federal Tax Policy for the 1980s*. Cambridge, Mass.: Ballinger.

Auten, G. E., & Clotfelter, C. T. (1982). Permanent Versus Transitory Tax Effects and the Realization of Capital Gains. *Quarterly Journal of Economics, 97* (November), 613–632.

Barnes, W. S. (Director) (1963). *World Tax Series: Taxation in the United States*. Harvard Law School, International Program in Taxation. Chicago, IL: Commerce Clearing House.

Bayly, R. A. (1902). *Succession Duties in Canada*. Toronto, Carswell.

Blakely, R. G. (1916). The New Revenue Act. *The American Economic Review, 6*(4), 837–850.

Blakely, R. G. (1917). The War Revenue Act of 1917. *The American Economic Review*, 7(4), 791–815.

Blakely, R. G. (1922). The Revenue Act of 1921. *The American Economic Review*, 12(1), 75–108.

Blakely, R. G. (1924). The Revenue Act of 1924. *The American Economic Review*, 14(3), 475–504.

Blakely, R. G., & Blakely, G. C. (1919). The Revenue Act of 1918. *The American Economic Review*, 9(2), 213–243.

Boadway, R. W., & Kitchen, H. M. (1984). *Canadian Tax Policy* (2nd ed.). Canadian Tax Paper No. 76. Toronto, Ontario: Canadian Tax Foundation.

Bradford, D. F. (1996). *Fundamental Issues in Consumption Taxation*. Washington, D.C.; The AEI Press.

Brinner, R. E. (1976). Inflation and the Definition of Taxable Personal Income. In: H. J. Aaron (Ed.), *Inflation and the Income Tax*. Washington, D.C.: The Brookings Institution.

Brozovsky, J., & Cataldo, A. J. (1993). The Marriage Tax Penalty: Inequities And Tax Planning Opportunities. *The Ohio CPA Journal*, (December), 22–26.

Brozovsky, J., & Cataldo, A. J. (1994). A Historical Analysis Of The 'Marriage Tax Penalty'. *Accounting Historians Journal*, 21(1), 163–187.

Bureau of the Census (1994). *Statistical Abstract of the United States 1995* (115 ed.). Washington, D.C.: U.S. Department of Commerce.

Bureau of the Census (1987). *Statistical Abstract of the United States 1988 (108th Edition)*. Washington, D.C.: U.S. Department of Commerce.

Burke, W. E. (1891). *Federal Finances or The Income of the United States*. Chicago, IL: F. J. Schulte & Co.

Burman, L. E., & Ricoy, P. D. (1997). Capital Gains and the People Who Realize Them. *National Tax Journal*, 50(3), 427–451.

Byrd, C., Chen, I., & Jacobs, M. (1994–1995). *Canadian Tax Principles*. Scarborough, Ontario: Prentice Hall Canada, Inc.

Byrd, C., Chen, I., & Jacobs, M. (1997–1998). *Canadian Tax Principles*. Scarborough, Ontario: Prentice Hall Canada, Inc.

Calegari, M. (1993). Changes to the Combined Income Tax and Social Security Tax Burden of Taxpayers who Realized Earned Income between 1960 and 1990. *Unpublished Working Paper* (University of Arizona, July 11).

Canadian Tax Foundation (CTF) (1967/1968–1994). *The National Finances*. Toronto, Canada.

Canadian Tax Foundation (CTF) (1990). *The National Finances*. Toronto, Canada.

Canadian Tax Foundation (CTF) (1995). *Finances of the Nation*. Toronto, Canada.

Carter, H., & Glick, P. C. (1970). *Marriage and Divorce: A Social and Economic Study*. Cambridge, MA: Harvard University Press.

Cataldo, A. J. (1995). The Earned Income Credit: Historical Predecessors and Contemporary Evolution. *The Accounting Historians Journal*, 22(1), 62–85.

Cataldo, A. J. (1996). Recent Developments in the Marriage Tax: A Comment and Decomposition. *National Tax Journal*, (December).

Cataldo, A. J., & Savage, A. A. (2000). The January Effect and Other Seasonal Anomalies: A Common Theoretical Framework. In: M. Epstein (Ed.), *Studies in Managerial Finance* (Vol. 9). Stamford, CT: JAI Press, Inc.

Clotfelter, C. T., & Steuerle, C. E. (1981). *Charitable Contributions included in How Taxes Affect Economic Behavior* (edited by H. J. Aaron & J. A. Pechman).

Collins, J. H. (1995). Discussion of An Empirical Investigation of Taxpayer Awareness of Marginal Tax Rates. *Journal of the American Taxation Association*, (Supplement), 60–61.

Congressional Budget Office (CBO) (1987). *Tax Policy for Pensions and Other Retirement Saving*. Washington, D.C.: U.S. Government Printing Office.

Congressional Budget Office (CBO) (1987). *The Changing Distribution of Federal Taxes: 1975–1990*. Washington, D.C.: U.S. Government Printing Office.

Congressional Budget Office (CBO) (1988). *How Capital Gains Tax Rates Affect Revenues: The Historical Evidence*. Washington, D.C.: U.S. Government Printing Office.

Cook, E. W., & O'Hare, J. F. (1987). Issues Relating to the Taxation of Capital Gains, *National Tax Journal*, 40 (September), 473–488.

Crum, R. E. (Ed.) (1992). *A Guide to Tax Research Databases*. Sarasota, FL: American Accounting Association.

David, M. (1968). Alternative Approaches to Capital Gains Taxation. *Studies of Government Finance*. Washington, D.C.: The Brookings Institution.

Dhaliwal, D., Trezevant, R., & Wang, S. (1992). Taxes, Investment-Related Tax Shields and Capital Structure. *Journal of the American Taxation Association*, (Spring), 1–21.

Enis, C. R. (Ed.) (1991). *A Guide to Tax Research Methodologies*. Sarasota, FL: American Accounting Association.

Ernst & Young International Ltd. (1991). *International Tax Rates*. New York, NY.

Evans, D., & Leighton, L. (1989). Some Empirical Aspects of Entrepreneurship. *American Economic Review*, (June), 519–535.

Evans, D., & Jovanovic, B. (1989). An Estimated Model of Entrepreneurial Choice Under Liquidity Constraints. *Journal of Political Economy*, (August), 808–827.

Feenberg, D. R., Mitrusi, A. W., & Poterba, J. M. (1997). *Distributional Effects of Adopting a National Retail Sales Tax*. Working Paper No. 5885 (January). Cambridge, MA: National Bureau of Economic Research.

Feenberg, D. R., & Rosen, H. S. (1995). Recent Developments in the Marriage Tax. *National Tax Journal*, 48(1), 91–101.

Feldstein, M. S., Slemrod, J. B., & Yitzhaki, S. (1984). The Effects of Taxation on the Selling of Corporate Stock and the Realization of Capital Gains. *Quarterly Journal of Economics*, 99 (February), 114–117.

Flemming, J. (1977). *Inflation*. Oxford University Press.

Flesher, T. K. (1985). *Statistics of Income Bibliography* (Arthur Young Foundation).

Fox, G. A. (1988). The Marriage Tax Penalty – 1980s Style. *The Ohio CPA Journal*, (Winter), 19–24.

Frischmann, P. J., Gupta, S., & Weber, G. J. (1996). New Evidence on Participation in Individual Retirement Accounts (IRAs). *Unpublished Working Paper* (February).

Gale, W. G. (1999). The Required Tax rate in a National Retail Sales Tax. *National Tax Journal, 52*(3), 443–457.

Gelardi, A. M. G. (1996). The Influence of Tax Law Changes on the Timing of Marriages: A Two Country Analysis. *National Tax Journal, 49*(1), 17–30.

Gentry, W. M., & Hubbard, R. G. (1997). *Fundamental Tax Reform and Corporate Finance*. Washington, D.C.: The AEI Press.

Gordon, R. H. (1992). Canada-U.S. Free Trade and Pressures for Tax Coordination. In: Shoven & Whalley (Eds), *Canada-U.S. Tax Comparisons* (pp. 75–96). Chicago & London: The University of Chicago Press.

Grasso, L. P., & Frischmann, P. J. (1992). Measuring Horizontal Equity: A Regression Approach. *The Journal of the American Taxation Association, 14* (Fall), 123–133.

Groves, H. M. (1963). Federal Tax Treatment of the Family. *Studies of Government Finance*. Washington, D.C.: The Brookings Institution.

Grubert, H., & Newlon, T. S. (1997). *Taxing Consumption in a Global Economy*. Washington, D.C.: The AEI Press.

Hansen, R. (1959). *The Tax Treatment of Family Income*. Ph.D. Thesis, University of Wisconsin.

Harvey, R. P., & Tempalski, J. (1997). The Individual AMT: Why it Matters. *National Tax Journal, 50*(3), 453–473.

Haugen, R. A., & Lakonishok, J. (1988). *The Incredible January Effect: The Stock Market's Unsolved Mystery*. Homewood, IL: Dow Jones-Irwin.

Hoffman, S. D., & Seidman, L. S. (1990). *The Earned Income Tax Credit: Antipoverty Effectiveness and Labor Market Effects*. Kalamazoo, MI: W. E. Upjohn Institute for Employment Research.

Holt, S. D. (1994). Effect of the 1993 Budget Act on the Advance Payment Option of the Earned Income Credit. *Tax Notes*, (February 7), 759–763.

Holtz-Eakin, D. (1995). Should Small Businesses Be Tax-Favored? *National Tax Journal*, (September), 387–395.

Holtz-Eakin, D., Joulfaian, D., & Rosen, H. S. (1994a). Entrepreneurial Decisions and Liquidity Constraints. *Rand Journal of Economics*, (Summer), 334–347.

Holtz-Eakin, D., Joulfaian, D., & Rosen, H. S. (1994b). Sticking it Out: Entrepreneurial Survival and Liquidity Constraints. *Journal of Political Economy*, (February), 53–75.

Hyman, D. N. (1987). *Public Finance: A Contemporary Application of Theory to Policy* (2nd ed.). New York, NY: Holt, Rinehart and Winston, Inc.

Hyman, D. N. (1999). *Public Finance: A Contemporary Application of Theory to Policy (Sixth Edition)*. Orlando, FL: The Dryden Press.

Ingram, J. A., & Monks, J. G. (1992). *Statistics for Business and Economics* (2nd ed.). Orlando, FL: Harcourt Brace Jovanovich.

Internal Revenue Service (IRS) (1951–1996). *Statistics of Income – Individuals.*

Internal Revenue Service (IRS) (1954, 1969 and 1976). *Statistics of Income – Estate Tax Returns.*

Internal Revenue Service (IRS) (1956). *Statistics of Income - Estate and Gift Tax Returns.*

Internal Revenue Service (IRS) (1958, 1960, 1962 and 1965). *Statistics of Income – Fiduciary, Gift, and Estate Tax Returns.*

Internal Revenue Service (IRS) (1992). 1989 Public Use Sample. *Statistics of Income – Individuals.*

Internal Revenue Service (IRS) (1995). *Estate Tax Returns, 1992–93, 14*(4). Statistics of Income (SOI) Bulletin.

Internal Revenue Service (IRS) (1996–1997). *Federal Taxation of Wealth Transfers, 1992–1995* (M. B. Eller), *16*(3). Statistics of Income (SOI) Bulletin.

Internal Revenue Service (IRS) (1997–1998). *Federal Estate Tax Returns Filed for Nonresident Aliens, 1995–1996* (M. B. Eller), *17*(3). Statistics of Income (SOI) Bulletin.

Internal Revenue Service (IRS) (1998–1999). *Projections of Returns To Be Filed in Calendar Years 1999–2005* (F. Zaffino), *18*(3). Statistics of Income (SOI) Bulletin.

Iyer, G. S., & Seetharaman, A. (1999). An Evaluation of Alternative Procedures for Measuring Horizontal Inequity. *Unpublished Working Paper.*

Iyer, G. S., Seetharaman, A., & Englebrecht, T. D. (1996). An Analysis of the Distributional Effects of Replacing the Progressive Income Tax with a Flat Tax. *Journal of Accounting and Public Policy, 15* (Summer), 83–110.

Jagolinzer, P., & Strefeler, J. M. (1986). Marital Status And The Taxes We Pay. *Journal of Accountancy,* (March), 68–77.

Johnson, C. H. (1996). Inefficiency Does Not Drive Out Inequity: Market Equilibrium & Tax Shelters. *Tax Notes,* (April 15), 377–387.

Joulfaian, D. (2000). A Quarter Century of Estate Tax Reforms. *National Tax Journal,* (September), 343–360.

Kamerman, S. B., & Kahn, A. J. (Eds) (1978). *Family Policy: Government and Families in Fourteen Countries*. New York: Columbia University Press.

Kirchheimer, B. (1993). EITC Gets High Marks, But Flaws Need Correcting. *Tax Notes,* (November), 521–523.

KixMiller, W., & Baar, A. R. (1917). *United States Income and War Tax Guide*. Chicago, IL: Commerce Clearing House.

Koenig, E. F. (1999). Achieving "Program Neutrality" Under a National Retail Sales Tax. *National Tax Journal, 52*(4), 683–697.

Kotlikoff, L. J. (1992). *Generational Accounting: Knowing Who Pays, and When, for What We Spend.* New York: The Free Press.

Laffie, L. S. (1998). Simplify! *Tax Advisor, 29*(6), 352.

Luttman, S., & Spindle, R. (1994). An Evaluation of the Revenue and Equity Effects of Converting Exemptions and Itemized Deductions to a Single Non-Refundable Credit. *The Journal of the American Taxation Association, 16* (Fall), 43–62.

Lyon, A. B., & Stell, J. L. (2000). Analysis of Current Social Security Reform Proposals. *National Tax Journal, 53*(3), 473–514.

Madeo, S. A., & Madeo, L. A. (1981). Some Evidence on the Equity Effects of the Minimum Tax on Individual Taxpayers. *National Tax Journal, 34* (December), 457–465.

Mariger, R. P. (1999). Social Security Privatization: What Are the Issues? *National Tax Journal, 52*(4), 783–802.

McGee, K. (1998). Capital Gains Taxation and New Firm Investment. *National Tax Journal, 51*(4), 653–673.

McIntyre, M. J. (1988). Rosen's Marriage Tax Computations: What Do They Mean? *National Tax Journal, 41*(2), 257–258.

McIntyre, M. J., & Steuerle, C. E. (1996). *Federal Tax Reform: A Family Perspective.* Washington, D.C.: The Finance Project.

Messere, K. (Ed.) (1998). *The Tax System in Industrialized Countries.* Oxford University Press.

Metcalf, C. E. (1972). *An Econometric Model of the Income Distribution.* Chicago, IL: Institute for Research on Poverty Monograph Series; Markham Publishing Company.

Miller, R. L. (1978). *Intermediate Microeconomics: Theory, Issues, and Applications.* McGraw-Hill, Inc.: New York.

Minarik, J. J. (1981). Capital Gains. In: H. J. Aaron & J. A. Pechman (Eds), *How Taxes Affect Economic Behavior.* Washington, D.C.: The Brookings Institution.

Mitchell, D. W. (1989). The Marriage Tax Penalty and Subsidy Under Tax Reform. *Eastern Economic Journal, XII*(2), 113–116.

MPL Communications, Inc. (1993). Family Composition Changing. *Canadian News Facts, 27*(11). Toronto: 4763.

Mullen, L. E. (1989). Who Should File Separate Returns? *The National Public Accountant,* (April), 30–33.

Musgrave, R. A., & Wilson, T. A. (1992). Reflections on Canada-U.S. Tax Differences: Two Views. In: J. B. Shoven & J. Whalley (Eds), *Canada-U.S. Tax Comparisons* (pp. 359–374). Chicago, IL: The University of Chicago Press.

National Revenue, Taxation (1936–1973). *Taxation Statistics.* Ottawa, Canada.

Neff, D. K. (1990). Married Women's Labor Supply And The Marriage Penalty. *Public Finance Quarterly, 18*(4), 420–432.

O'Neil, C. J., & Nelsestuen, L. B. (1994). The Earned Income Credit: The Need for a Wealth Restriction for Eligibility Determination. *Tax Notes,* (May 30), 1189–1201.

Pearlman, R. A. (1996). Transition Issues in Moving to a Consumption Tax: A Tax Lawyer's Perspective. In: H. J. Aaron & W. G. Gale (Eds), *Economic Effects of Fundamental Tax Reform*. Washington, D.C.: Brookings Institution Press.

Pechman, J. A. (1987). *Federal Tax Policy* (5th ed.). Studies of Government Finance. Washington, D.C.: The Brookings Institution.

Pechman, J. A. (1971). *Federal Tax Policy* (Rev. ed.). Studies of Government Finance. Washington, D.C.: The Brookings Institution.

Pechman, J. A. (1985). *Who Paid the Taxes, 1966–85*. Washington, D.C.: The Brookings Institution.

Pechman, J. A., & Engelhardt, G. V. (1990). The Income Tax Treatment of the Family: An International Perspective. *National Tax Journal, 43*(1), 1–22.

Pechman, J. A., & Okner, B. A. (1974). *Who Bears the Tax Burden*. Washington, D.C.: The Brookings Institution.

Perry, J. H. (1955). *Taxes, Tariffs, & Subsidies: A History of Canadian Fiscal Development*. Canadian Tax Foundation, University of Toronto Press.

Phillips, L. C., & Previts, G. J. (1983). Tax Reform: What are the Issues? *Journal of Accountancy, 155*(5), 64–74.

Pierce, B. J. (1989). Homeowner Preferences: The Equity and Revenue Effects of Proposed Changes in the Status Quo. *The Journal of the American Taxation Association, 10* (Spring), 54–67.

Plotnick, R. (1984). *A Comparison of Measures of Horizontal Inequity Using Alternative Measures of Well-Being*. Discussion Paper 752–84, Institute for Research on Poverty; University of Wisconsin-Madison.

Pollack, S. D. (1996). *The Failure of U.S. Tax Policy: Revenue and Politics*. University Park, PA: The Pennsylvania State University Press.

Research Institute of America (RIA) (1997). *1998 RIA Federal Tax Handbook*.

Research Institute of America (RIA) (1995). *1996 RIA Federal Tax Handbook*.

Revenue Canada, Taxation (1974–1998). *Taxation Statistics*. Ottawa, Canada.

Ricketts, R. C. (1991). Social Security Growth Versus Income Tax Reform: An Analysis of Progressivity and Horizontal Equity in the Federal Tax System in the 1980s. *The Journal of the American Taxation Association*, (Spring), 34–50.

Ricketts, R. C. (1990). Social Security Growth Versus Income Tax Reform: An Analysis of Progressivity and Horizontal Equity in the Federal Tax System in the 1980s. *The Journal of the American Taxation Association, 11* (Spring), 34–50.

Rignano, E. (1924). *The Social Significance of the Inheritance Tax*. New York: Alfred A. Knopf.

Rosen, H. S. (1997). The Way We Were (And Are): Changes in Public Finance and its Textbooks. *National Tax Journal (50th Anniversary Issue), 50*(4), 719–730.

Rosen, H. S. (1988). Thinking About The Tax Consequences Of Marriage. *National Tax Journal, XLI*(2), 259–260.

Rosen, H. S. (1987). The Marriage Tax Is Down But Not Out. *National Tax Journal, 40*(4), 567–575.

Ruffin, R. J., & Gregory, P. R. (1993). *Principles of Economics* (5th ed.). New York: HarperCollins College Publishers.

Rupert, T. J., & Fischer, C. M. (1995). An Empirical Investigation of Taxpayer Awareness of Marginal Tax Rates. *Journal of the American Taxation Association, 17* (Supplement), 36–59.

Scholes, M. S., & Wolfson, M. A. (1992). *Taxes and Business Strategy: A Planning Approach.* Englewood Cliffs, NJ: Prentice-Hall.

Scholz, J. K. (1994). The Earned Income Tax Credit: Participation, Compliance, and Antipoverty Effectiveness. *National Tax Journal,* (March), 63–87.

Seldon, G. E. (1920). Year-end selling a feature. *Magazine of Wall Street,* (December 25), 230–231.

Seltzer, L. H. (1959). The Place of the Personal Exemption in the Present-Day Income Tax, testimony submitted to the House Committee on Ways and Means. *Tax Revision Compendium, 1* (November), 508–509.

Sennholz, H. F. (1976). *Death and Taxes.* Washington, D.C.: The Heritage Foundation.

Shlaes, A. (1999). *The Greedy Hand: How Taxes Drive Americans Crazy and What to Do About It.* Random House, NY.

Shoven, J. B., & Whalley, J. (Eds) (1992). *Canada-U.S. Tax Comparisons.* NBER Project Report. Chicago & London: University of Chicago Press.

Sjoquist, D. L., & Walker, M. B. (1995). The Marriage Tax and Timing of Marriage. *National Tax Journal, 48*(4), 547–558.

Slemrod, J. (Ed.) (1994). *Tax Progressivity and Income Inequality.* Cambridge University Press.

Smith, A. (1977). *The Wealth of Nations.* The Everyman's Library Edition (J. M. Dent & Sons Ltd.).

Social Security Administration (1987). Social Security Programs in the United States. *Social Security Bulletin,* (April).

Statistics Canada (various years). *The Canada Yearbook.* Ottawa, Canada.

Steuerle, G. (1993). Economic Perspective: Measuring the Impact of Job Subsidies: Is There a Double Standard? *Tax Notes,* (November 15), 865–866.

Strayer, P. J. (1939). *The Taxation of Small Incomes: Social, Revenue, and Administrative Aspects.* New York: The Ronald Press Company.

Strefeler, J. M. (1982). The Tax Penalty On Marriage: An Odious Wedding Gift. *The Woman CPA,* (October), 5–10.

Tanzi, V. (1976). Adjusting Personal Income Taxes for Inflation: The Foreign Experience. In: H. J. Aaron (Ed.), *Inflation and the Income Tax.* Washington, D.C.: The Brookings Institution.

Treff, K., & Cook, T. (Eds) (1995). *Finances of the Nation.* Toronto: Canadian Tax Foundation.

Wagner, R. E. (1977). *Inheritance and the State: Tax Principles for a Free and Prosperous Commonwealth.* Washington, D.C.: American Enterprise Institute for Public Policy Research.

Wall Street Journal (WSJ) (1999). Estate-Tax Relief is Coming in Less Than Two Months. *Tax Report*, (Wednesday, November 3), 1.

Wall Street Journal (WSJ) (1999). Careful Timing of a Wedding May Save a Bundle at *Tax Time*. *Tax Report*, (Wednesday, November 17), 1.

Wall Street Journal (WSJ) (2000). Margin Calls Rise. *Adding Pressure to Sell-Off*, (Monday, April 17), A3.

Wetzler, J. W. (1977). Capital Gains and Losses. In: J. A. Pechman (Ed.), *Comprehensive Income Taxation*. Washington, D.C.: The Brookings Institution.

Willis, H. P. (1914). The Federal Reserve Act. *The American Economic Review, 4*(1), 1–24.

Yin, G. K., & Forman, J. B. (1993). Redesigning the Earned Income Tax Credit Program to Provide More Effective Assistance for the Working Poor. *Tax Notes*, (May 17), 951–960.

Zodrow, G. R. (1999). The Sales Tax, the VAT, and Taxes in Between – or, Is the Only Good NRST a "VAT in Drag"? *National Tax Journal, 52*(3), 429–442.

AUTHOR AND SUBJECT INDEX

Where appendix (A.), footnote or endnote (n.), table (t.), figure (f.), and glossary of terms (g.) is also indicated and precedes or follows page numbers.